T0198037

Beatlemania

JOHNS HOPKINS INTRODUCTORY STUDIES IN THE HISTORY OF TECHNOLOGY

BEATLEMANIA

Technology, Business, and Teen Culture in Cold War America

ANDRÉ MILLARD

The Johns Hopkins University Press
Baltimore

The Johns Hopkins University Press
2715 North Charles Street
Baltimore, Maryland 21218-4363
www.press.jhu.edu

Library of Congress Cataloging-in-Publication Data
Millard, A. J.
   Beatlemania : technology, business, and teen culture in cold war America / André Millard.
      p. cm. — (Johns Hopkins studies in the history of technology)
   ISBN-13: 978-1-4214-0524-7 (hdbk. : alk. paper)
   ISBN-13: 978-1-4214-0525-4 (pbk. : alk. paper)
   ISBN-13: 978-1-4214-0627-5 (electronic)
   ISBN-10: 1-4214-0524-5 (hdbk. : alk. paper)
   ISBN-10: 1-4214-0525-3 (pbk. : alk. paper)
   ISBN-10: 1-4214-0627-6 (pbk. : electronic)
      1. Beatles. 2. Rock musicians—England—Biography. 3. Rock music—Marketing.
   4. Branding (Marketing)—Social aspects. I. Title.
   ML421.B4M55 2012
   782.42166092′2—dc23   2011042374

A catalog record for this book is available from the British Library.

*Special discounts are available for bulk purchases of this book. For more information,
please contact Special Sales at 410-516-6936 or specialsales@press.jhu.edu.*

The Johns Hopkins University Press uses environmentally friendly book materials,
including recycled text paper that is composed of at least 30 percent post-consumer
waste, whenever possible.

Contents

Preface

This is a book about a social phenomenon rather than a history or an appreciation of a great band. It concentrates on the relationship of the Beatles with the fans and on the historical context that made Beatlemania one of the best remembered cultural events of the 1960s. There are so many books about the music that it seems redundant to add any more praise to the Beatles' artistic reputation. In the preface to Walter Everett's excellent study, *The Beatles as Musicians* (Oxford University Press, 2001), he says that they are "the most important force in twentieth-century music." I agree. But might other factors outside their beloved music explain why they were such a force? Is there another way to understand Beatlemania without assigning it to the power of the songs and the creativity of the musicians?

Rather than examine what John, Paul, George, and Ringo put into the music, this book looks at what their fans took out of it: what they heard when they listened to the records, and what they experienced when they went to the shows. It charts their interaction with an image as well as with the music. It seeks to explain the unprecedented popularity of the band by putting it into a broader historical context, specifically the transatlantic diffusion of culture and the operation of multinational business organizations.

The forces that created Beatlemania gathered momentum in the 1950s and 1960s, but the longer waves of technological development and cultural exchange started back in the nineteenth century and culminated at a special moment in the relationship of the United States and Great Britain. Beatlemania emerged from the convergence of economic, technological, and demographic forces. A major focus of this book is the technology, not just the machines but the ways in which the Beatles, their fans, and the entertainment business used technology to connect with one another.

I consider the Beatles' music as a commercial product rather than a work of art and seek its origins in the interchange of ideas, skilled technicians, and sound recordings across the Atlantic. This study looks for Beatlemania at the intersection of commerce and entertainment, rather than in the realms of psychology and sociology. It develops a narrative based on the collective memories of the fans, and uses their hopes and dreams to fill in the spaces between the lines of those wonderful songs. Beatlemania captured a magic moment when four musicians entered the lives of millions of teenagers and wrought the kinds of changes that made the sixties the sixties. This book is devoted to that time and to those cherished memories that made the years of Beatlemania so meaningful.

Much of my thinking about the Beatles has been influenced by teaching classes on American popular culture, the music industry, and the British Invasion. Over the years my students have been a sounding board for my ideas and have shared their interpretations of the band and the music with me. I want to thank them all, especially graduate students Allen Hyde, Ken King, Scott Reynolds, Russell Richey, James K. Turnipseed, and Amanda Cody. My colleagues Brian Steele, John Van Sant, and Dr. John Sims gave me the benefit of their expert knowledge. My good friends in the Birmingham Record Club, especially Charles Bailey and Gus Liberto, shared not only their wonderful record collections with me, but also their own experiences of Beatlemania. These guys lived it, and it has been so helpful to talk with the people who were there in person. In England, the staffs of the Liverpool Central Library and the British Library newspaper division were most helpful, as were my informants on the English music scene, Ted Scannell and Mike Wood, and Frank Seltier in Germany. Nathan Pendlebury and Kay Jones of Liverpool Museums helped me with the photographs, and Margaret Roberts of Peter Kaye photographers shared her wonderful recollections of sixties Liverpool with me. Michele Forman helped to prepare the manuscript. A final word of thanks to my editor at the Johns Hopkins University Press, Bob Brugger, whose good humor and judgment I have relied on for all these years. Thanks.

Beatlemania

CHAPTER 1

THE RECORD

The world of the Beatles was defined by records. Recorded sound was the inspiration to form a group, and over time it became their vocation—they thought of themselves primarily as recording artists rather than performers or entertainers. Recordings transformed four individual lives in Liverpool and brought them together into the Beatles. All of them enjoyed listening to records before they decided to become musicians. They started collecting records as a hobby, and it became a passion. Each of them had an extensive record collection by which they mutually discovered not only the sort of music that spoke to them, but also the musicians they wanted to be.

The common ground of recordings brought John Lennon and Paul McCartney together that sunny afternoon in July 1957 when they first encountered one another at a church fete—a meeting as important in the history of popular music as the day that Elvis Presley walked into Sam Phillips' Sun Recording Service in Memphis, Tennessee, in August 1954. John's band, the Quarry Men, was playing one of its first gigs, and Paul was in the audience. They were introduced by a mutual friend, and their first exchanges, as they sized each other up, were about the records they knew. If Paul hadn't heard the Del-Vikings "Come Go with Me," which was part of the Quarry Men's repertoire, or listened to Eddie Cochran's "Twenty Flight Rock" and learned all the words, who knows if he and John would have hit it off.[1]

When the Beatles were growing up in Liverpool, records were still an expensive luxury. Buying one was an event for teenagers, often done with close friends to spread around the warm glow of consumption. Most participants in Beatlemania can tell you when they first heard or saw the Beatles, but they can also recall exactly when and where they bought the records. Once safe in the collection, records were played over and over again into the night, with fans not just hearing, but absorbing them, letting the words and music sink in. In the Beatles' Liverpool, records were borrowed, swapped, resold, and passed around. They were the glue that brought groups of schoolboys together, in rituals of purchase (or shoplifting), listening, discussion, and enlightenment. Tony Bramwell, a close boyhood friend of George and Paul, emphasized how important that revolving disc was to them: "Seven inches of black plastic with a hole in the middle. Life, magic."[2] In the status-hungry world of English and American teenagers, the record collection said a lot about the owner. It was the sixties form of Facebook—an assessment of your outlook and a measure of cool. Speaking of American rhythm and blues (R&B) records, John Lennon remembered: "We felt very exclusive and underground in Liverpool listening to all those old time records." All the Beatles and their manager were proud of their collections. Brian Epstein's will, written in 1956, decided the disposition of his "artistic possessions," which included "records, magazines, theatrical programs."[3]

By mutual consent the Beatles and their boyhood friends believed that Liverpool was the largest and most enthusiastic audience for American R&B records in the country, and no doubt these recordings were the most valued. They were the Rosetta stone of the new youth music of the 1950s—the treasured, historic discs that brought the sound of Memphis, Chicago, and New Orleans to them with the sacred message of rock'n'roll. One of them was "Heartbreak Hotel," which introduced Elvis Presley to European listeners and usually gets a lot of the credit for establishing rock'n'roll over the Atlantic. The effect of this record on youngsters like John Lennon was immediate and astounding: "That was the conversion. I kind of dropped everything," he said later. Although other American singers had impressed Lennon, there was nothing to compare with the impact of "Heartbreak Hotel." The eerie, reverb-laden vocals, and the melodrama in Elvis's delivery, put it apart from all the others. George Harrison remembered the seismic impact of its release: "When a record came along like 'Heartbreak Hotel' it was so amazing."[4] Both John and George would agree that one record could change your life. American records formed the basis of the Beatles' musical education, and they continued to play an important role in the band's development. Recordings by Bob Dylan and the Beach Boys, such as *The Freewheelin' Bob Dylan* and *Pet Sounds*, exerted a powerful influence on Lennon and McCartney's music.[5]

Liverpudlians (the name given to its people) felt a special affinity with America and its music. The city on the river Mersey looked across the Atlantic, and the Merseyside bands played American music. Acquiring blues and jazz records took the collector deeper and deeper into African American music and strengthened the imaginary ties between young men in Liverpool and the mythological South. Paul said of the Beatles, "We were all very interested in American music and all

had the same records," and it was their common admiration of these recordings that was the foundation of their group. The United States was the home
of "all the music we loved," and this was why they were so excited to be touring
America. In fits of anxiety over the prospects of their first American tour, both
John and Paul expressed grave doubts about the outcome, but then cheered up
with the prospect of all the records they could buy. Each trip the Mersey bands
made to America was seized upon as an opportunity buy records, and they always
returned, in the words of Billy J. Kramer, "with loads of albums."[6]

Recordings were the basis of their musical education in the total absence of
any reading skills and formal instruction. A copy of guitarist Bert Weedon's in-
structional book *Play in a Day* could be found in their bedrooms and basements,
but like every other guitar group, the Beatles learned from listening to records.
They brought them to their first practices as a band and continued to bring them
into recording sessions as they matured. In the crowded beat scene of Liverpool
in the early 1960s, the key to building a repertoire was to find obscure R&B
and country songs—the B sides, the "shots of rhythm and blues" that Paul sang
about. Record collecting was part of being in a beat group.

While Paul McCartney listened to Chan Romero's "Hippy Hippy Shake" in
a record shop, his friend Prem Willis-Pitts was in an adjacent booth enjoying
the same song. This was a hot new record, raw and explosive, with a scorching
guitar break. Both bought the record that day with a mind to copy it on stage, but
someone from another band, the Swinging Blue Jeans, beat them to it, and their
version became a hit.[7] The Beatles faced some tough competition in the Mersey
Beat scene in the early 1960s, and the way to get an advantage over excellent
groups like the Searchers or Big Three was to find new songs to copy.

Making a record was central to the business plan of the Beatles and all the
groups around them. Records traced the path to success, from the first amateur
recordings on a do-it-yourself acetate disc to the gold record trophies on the walls.
Ringo expressed a common sentiment that "you'd sell your soul for a record," and
Gerry Marsden (of Gerry and the Pacemakers) justified being in a group because
"at least we can tell our kids we made a record."[8] Tony Bramwell recalls George's
pride in the first "single" made by the Quarry Men, a one-off, acetate, home
recording made for one pound, but "this was a real record, and scratchy or not,
it sounded authentic, like the beginning of something."[9] Although the Beatles
were drawn to film making as a measure of creativity, their personal evaluation
of success was always in terms of recordings: "We've only ever gone with record
sales. When we stop selling records, we'll probably pack it in. We came into this
business . . . to sell records."[10] The triumph of the Beatles was first and foremost

the triumph of the record—its fidelity, price, packaging, and the part it played in peoples' lives. The Beatles saw their musical heritage in terms of recordings, and today these pieces of vinyl are the core artifact of their story.

RECORDS AND HISTORY

The centrality of acquiring records in teenage culture made the record shop a place of some significance for the Beatles and their peers in Liverpool. They came to browse and interact with one another in the updated and electrified version of the eighteenth-century coffee shop. They were not drawn to Brian Epstein and his North End Music Shop (NEMS) by fate; it was a very special store—one of the largest outside the capital—and it catered to the esoteric tastes of its knowledgeable and demanding customers.

If he had not managed the Beatles, Brian Epstein would be remembered in the record business as an innovative retailer. True to his theatrical background, Epstein took pains to make his store appear glamorous, and baby boom Liverpudlians remember his displays with affection. His system of stock control was copied by London stores. He had a policy of ordering any record in print for a customer, which in any accountant's estimation would be a waste of time and money. But this commitment to seek out any record made a statement to the customers. The musicians made a point of checking out the box of new releases situated next to the cash register. On Saturday afternoons the shop would be packed with them, including the Beatles ("they were always in there"), listening to records in the booth opposite the counter, browsing through the piles of singles, admiring the displays of LP covers, or just socializing with their friends.[11]

One record had a special place in the history of the Beatles because it brought them together with Brian Epstein. Their first commercial record, in which they backed Tony Sheridan on "My Bonnie," was released as a single in July 1961 in Germany on the Polydor label. Soon people were coming into NEMS to ask for it, no doubt directed there by deejay Bob Wooler of the nearby Cavern Club, where the Beatles had established themselves. Epstein's interest was stimulated by a request for the record by Raymond Jones, and his subsequent visit to the Cavern led to the historic encounter with the Beatles. It makes for a good story and reflects well on the business acumen of Epstein.

The English tabloid newspaper the *Daily Mirror* traced the whole fabulous history of the Beatles back to the moment when Epstein went out of his way to please a single customer. Alistair Taylor, a NEMS employee, initiated the order, and he claims that he made up the name to get the record into the accounting system and justify ordering it. But devoted Beatles historians have actually located Mr. Jones, and he claims that he got an early listen of the German record and then ordered it from NEMS.[12] It doesn't really matter who ordered the record because Epstein *must* have known about the band; he was associated with the newly created *Mersey Beat* newspaper, which regularly printed photos and articles about the group. They played the Cavern regularly, and it was only a short distance from the shop, where they were regular customers, well known to Taylor and the rest of the staff. So why invent a story that could not be true? It maintains the centrality of the record in the story of the Beatles.

The hit record provides the markers in the history of popular music. A record contract was proof of the professionalism of both musicians and manager. Epstein's close links (via his retail business) with the record industry was the deciding factor in the Beatles' acceptance of his management offer. (That and the big car he drove.) As soon as they signed the contract, he used his connections with the Big Two record companies in England—Decca and Electric and Musical Industries (EMI)—to arrange auditions, and before the end of 1961, Decca had sent scouts to the Cavern and booked a studio for the Beatles to play for them.

Thus far the story follows the standard transcript of rock'n'roll stardom: a chance encounter, a record contract, and then a hit record that lights up the telephones at the radio station as the kids phone in. Yet at this crucial point, the plan fell through because Decca famously rejected them. Again the record shop enters the story, in this case the flagship HMV (His Master's Voice) store in central London, where Epstein had gone to make acetates of the Beatles' demonstration (demo) tapes in his last-ditch effort to get the band signed. In those days a record store did more than sell records. Some had booths to record a song (such as the one Elvis Presley used to first record his voice), and others, like the HMV store, had better equipment and a handsome reception area.

The demos were recorded on acetate discs, the "instant" but fragile recordings that could be recorded on only once and then deteriorated with every play. But the acetates were handy and could be played on the small players that sat on the desks of every manager, producer, venue booker, and agent. The engineer who transferred the Beatles tapes to acetate discs was impressed, and his call to Ardmore and Beechwood, the publishers of EMI's music, gave Epstein the chance to make his pitch to George Martin, and he set up the fateful Parlophone session for June 1962. This was such an important event in their lives that one of them saved the historic telegram by which Brian conveyed the good news: CONGRATULATIONS BOYS. EMI REQUESTS RECORDING SESSION. PLEASE REHEARSE NEW MATERIAL.[13]

COUNTING IN RECORDS

In the popular music industry, the record was both talisman and signifier, incorporating the magic of creativity in its grooves and acting as the trophy of success. In this business, you counted in terms of record sales. The sales charts, the *Billboard* top sellers, the hit parade, as it was called in England, provided the data that determined the operation of record and radio businesses. The charts were how businessmen kept score, and professionals who played the game on both sides of the Atlantic examined them daily. The American charts were routinely published in the United Kingdom, and the Top 30 of the British *New Musical Express* (NME) was published in *Cash Box* and *Billboard*. A band was invisible until a record was released, and then the transatlantic network of charts started to track the artists. In this way Tony Sheridan, the Beatles, and their record "My Bonnie" entered the system in late 1961, and their names started to appear in the musical press.[14]

Chart-topping records form a narrative of Beatlemania, beginning with the ascent of "I Want to Hold Your Hand" in the United States in January 1964 and continuing with all the other record-breaking statistics: the fastest selling single, the longest stay in the Top 10, the biggest jump up the charts, the largest number preordered before the release, and the number of records in the Top 5, 10, and 100 published by *Billboard*. These statistics provide the skeleton for the story. Record sales were central to the superlatives generated by, and supportive of, Beatlemania—as if selling records was part of some other, even more significant, event that revealed the magnitude of what was unfolding in North America. Of all the millions of Beatlemania memorabilia, the records are the most evocative and revealing.

All along the way, the records tell the story. Each contains a bit of the Beatles and their history. The long-player *Meet the Beatles!* is a cherished memory of Beatlemania and the beginning of the affair for many. *Sgt Pepper's Lonely Hearts Club Band* stands as an iconic artifact of the swinging sixties, especially Peter Blake's famous assembly of images for the cover. *The Beatles* (known as *The White Album*) tells its own sad story of four musicians going their own ways.

The prehistory of Beatlemania begins with the 45 rpm singles released in the UK by Parlophone (a label owned by the EMI company) before Americans had even heard of the band. "Love Me Do" and "Please Please Me," issued in the fall of 1962, came in dark green paper sleeves and cost six shillings and three pence. In the center of the disc was the label in old-fashioned black with dignified uppercase lettering surmounted by the Parlophone brand—an ironic cursive L that is also used in the UK to signify units of pounds sterling. (Was this a portent of the profits to follow and the first secret message embedded in a Beatles record?)

The Parlophone single "She Loves You" reached Number 1 in the UK in September 1963, and its "yeah yeah yeahs" announced the arrival of Beatlemania. The sales figures were astounding: nearly 750,000 copies sold in the first month of release made it the fastest selling record that had ever been issued in the UK. It stayed at Number 1 for a month and remains the Beatles' biggest selling single; but most important, it articulated the joy of the fans amid the upward trajectory of the band. When the Cavern faithful heard that the record had reached the top of the charts, however, they were filled with despair because they knew that the Beatles were now too big to stay in their hometown. The fans were right, for the move to London was under way. "She Loves You" concluded the Liverpool chapter of the story.

The American archive of the British Invasion begins with the Beatles' singles issued by Capitol. These were the hard currency of Beatlemania, conferring fan status on their owners and spreading the magic of the band's music. They came with the brightly colored orange and yellow Capitol label, which featured the swirl, a circular wavelike design that had a lot more American pizzazz than the staid Parlophone centerpieces. For the great majority of American fans, Beatlemania began with "I Want to Hold Your Hand." Joining "Beatleworld" or becoming a "Beatlemaniac" required more than listening to these recordings—you had to own a copy: "I went to school on the Friday before the *Ed Sullivan Show*, and *every single person* on the bus . . . had the 45 . . . It was like *the* thing to walk around carrying that single."[15] So significant were these records that they were eagerly displayed and shared with special friends. When American teenagers bought "I Want to Hold Your Hand," they were getting a lot more than a sound recording; the record gave them status and secured inclusion in a most fashionable group. Buying the records was the necessary part of being a fan. But being the first to buy the latest release, or getting hold of a rare recording, conveyed even more prestige: "I was always the first kid on the block to get whatever the newest album was . . . It was a cool place to be."[16] Beatlemania lives on today in those cherished pieces of vinyl. They are still preserved in boxes in the basement or in gilded frames in the living room. Such is the value placed on them that there is a booming trade in counterfeits today. According to Bruce Spizer, *Introducing the Beatles* is the most counterfeited album of all time. Those perfectly preserved copies offered for sale at record shows all over the country are more than likely fakes.[17]

When the Beatles first started to collect records, the standard format was the

78 rpm shellac disc, a large, 10" diameter, heavy disc that had been in use for much of the twentieth century. After World War II, the American Columbia and RCA companies developed new discs made out of vinyl that could hold smaller (micro) grooves and much more music. In 1948 Columbia introduced its 12" long-player, revolving at 33⅓ rpm, and RCA a smaller 7" disc that revolved at 45 rpm and had shorter play time. These were very slowly adopted in the United Kingdom, and pop singles were still released on shellac discs throughout the 1950s. RCA persisted with its 45 rpm format and in 1952 brought out an extended play (EP) 7" disc that could hold around eight minutes of sound, enough for three or four short songs. The EP was hardly used in the United States, but in England it became a popular format that sat between the single and the long-playing album (LP). Cash-strapped English teenagers who could not afford the LPs (which sold at over one pound, ten shillings) liked the intermediate EP, which the Beatles played their part in popularizing. *The Beatles Hits,* with four songs on it, came out in 1963.

LPs were as important as the singles in introducing the Beatles to their public. The aptly named Capitol long-player *Meet the Beatles!* brought the band's music to millions of Americans in 1964. From *Please Please Me,* the best document of them as a hard-rocking band, to *Rubber Soul,* and then to their triumphant *Sgt Pepper,* LPs were significant markers of the ascent of the band as creative artists. After Beatlemania had subsided in 1966, the Beatles' progress as musicians was measured in LPs, which marked each upward step in their music and reputation.

Long-playing albums also started the Beatles on their romance with film making. Albums were the perfect vehicle for soundtracks from Broadway shows, and LPs like *My Fair Lady* and *The Sound of Music* broke many sales records. Rock and pop soundtracks from films were also a profitable line for record companies and film studios. The United Artists (UA) studio approached the Beatles management with a film offer not because they thought the band members could act but because they wanted an opportunity to sell their music in America. Like many other concerns in the film industry, UA had formed its own record division in 1958 to exploit the growing market for rock'n'roll.

EMI owned the Beatles' contract and issued their records on its Parlophone label, and it had made an agreement with Capitol to market Beatles records in the United States but had inexplicably left soundtrack albums out of the contract, thus leaving a loophole for another company to release Beatles soundtrack albums in the United States. United Artists signed the band to give them the right to release the music from the films through their record company. The integrated structure of these global entertainment companies allowed them to exploit the synergies of sound and image, and UA saw in the Beatles a wonderful opportunity to develop its record division. It was not UA's film production arm in London that first noticed the Beatles, but rather the Artist and Repertoire (A&R) men of the record division who pushed the company to contact Brian Epstein. As David Picker, head of production, explained: "We made the deal for one reason . . . we were expanding our music recording company so we were going to get publishing rights and a soundtrack album. I'd never even see them perform."[18]

THE FACES ON THE COVER

The LP brought more than just forty minutes of music. For the first half of the twentieth century, record companies had resisted the idea of illustrated covers on their products as an unnecessary expense. Picture sleeves had been introduced in the United States in the late 1950s on selected singles, mainly of attractive young male crooners, and the record companies found that a colored cover had a positive effect on sales. Thus Capitol released "I Want to Hold Your Hand" in an illustrated sleeve showing the four Beatles in their trademark collarless jackets. The success of EPs in the UK, which always had picture covers, and the universal popularity of the LP with its square cardboard cover, pushed the record companies to establish art departments to supply a stream of promotional images. The digital CD in its jewel case is the final remnant of that old-fashioned system of selling pre-recorded music in a product you can touch and examine for visual clues. Fifty years ago the album cover was just coming into prominence, and the Beatles were the first to take advantage of this development.

If we take each of the four Beatles as a representative example of record buyers of the 1960s, we know that they examined LP covers carefully, examining the guitars and other gear shown, digesting the information on the liner notes to see who wrote what songs. They memorized the images as well as the music on their favorite records. When it was their turn to put together an album, they put the same care and forethought into the picture that went on the cover because the look was important to them. The first English LP, *Please Please Me*, is a typical example of the cover art of the times: a group shot of the members close together in full lurid color, framed against a bland but evocative background and shot at an angle—a diagonal—to make the image a bit more interesting. That was why guitar bands often posed on stairs for their cover photographs. It was customary in the UK to leave a small band of white on the top of the cover for printing the title, artist, and information on the format—mono or stereo.

THE BEATLES

I SAW HER STANDING THERE
I WANT TO HOLD YOUR HAND

5112

This cover image took America by storm and helped ignite Beatlemania. Released a few weeks before the Beatles arrived in the United States, "I Want to Hold Your Hand" became the hit record that brought their music to America. The B side was "I Saw Her Standing There," but Capitol Records, hedging its bet, reversed the order in some releases and put "I Saw Her Standing There" on the A side. The cover shows the image that Brian Epstein created for his boys—collarless suits and the unique hair—but his eagle eye missed the cigarette in Paul's hand. Capitol airbrushed it out in some of the re-releases. (© Apple Corps., Ltd.)

The Beatles / 1962–1966

On this collection of Beatles hits, EMI used the same picture from the cover of the Beatles first LP *Please Please Me*. In collecting later hits (1966–1969) the firm replaced the inexperienced boy band in this picture with a photograph of the Beatles as they had become—four of England's most famous hippies. The idea was to compare the present with the past, but the most potent comparison turned out to be the album that followed *Please Please Me*. (© Apple Corps., Ltd.)

The cover of their second album, *With the Beatles,* broke from this practice. The Beatles had a lot more control of the cover image than anyone else before them. In the record business, a company employee decided on the cover art just as the music was chosen by their A&R man (who matched the songs to the artist) and recorded by their engineers in the company's studio. But the Beatles were very interested in the way they were depicted and wrested control of this process from EMI. They were pioneers in determining the artwork on their records. Art director George Osaki remembered: "When I started, the artist never looked at the album covers . . . The Beatles changed all that because they made me show them what I was going to do for the album cover."[19]

Brian Epstein and the Beatles picked photographer Robert Freeman to shoot the cover image because they were impressed with his pictures of jazz musicians. At that time black-and-white images were only used for jazz records, in contrast to the gaudy colors of pop albums. The photograph chosen for the cover of *With the Beatles* is reminiscent of the pictures that their friend Astrid Kirchherr took of the group while they were in Hamburg: stark, grainy, and moody, with the faces illuminated from the light source coming in from the side, leaving them half in the shadow. Kirchherr recalled that her idea was to represent the boys "looking so rough and somehow so full of wisdom," and this sums up perfectly the visual style of band portraits that still dominates cover art today.[20] Freeman arranged John, Paul, and George in a line and had Ringo kneel down, which organized the four faces into the rectangular shape of the cover, but in the printing, the images appeared slightly darker than expected, which made the faces appear disembodied from the bodies and background and gave the images an unusual intensity as they seemed to float within the space of the frame. (Because of the slight difference in the printing process, the black on the cover of the American record had a blue tint.)

This photograph now has iconic status, and the first impression of those who bought the album was one of shock—nobody had seen an LP cover like this before. It is no wonder the cover of *Meet the Beatles!* (the name given to the American LP by Capitol) made an immediate impression on the fans as something new and stylish. They had noticed that the band looked different from the other pop musicians they had seen on television. The cover for *Meet the Beatles!* underlined the European chic of the Beatles and their otherness. The band continued to be innovative in their cover art, which provided visual clues to the rapid progression from boy band to artistes, from followers to leaders in style. The "White" album was a deliberate step back to minimalism at a time when crowded, brightly col-

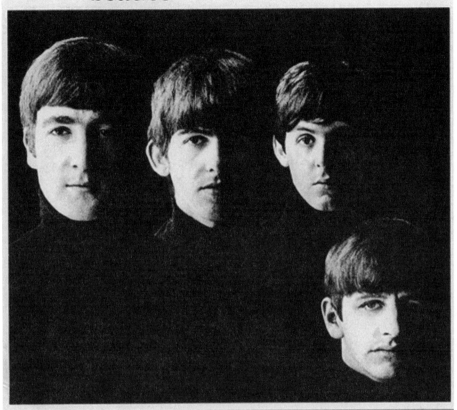

Robert Freeman took this iconic image of the Beatles, which marked a dramatic departure from the style of record covers used for pop bands. The printing of the cover turned out a little darker than expected, and the faces of the four musicians seemed to float disembodied from the background. Record buyers had never seen a cover like this before. It marked the rise of the Beatles' style, creativity, and power within the record company. (Courtesy Apple Corps.)

ored psychedelic images were the fashion after *Sgt Pepper's Lonely Hearts Club Band* had established the trend.

The LP cover is also an artifact of the changing relationship between record company and artists. It took George Martin's intervention to convince EMI to approve the cover of *With the Beatles*, which marks a significant step in the protection of artistic autonomy that would be one of the Beatles' most important contributions to popular music. Yet it was still a compromise because EMI prevailed over the question of removing the upper white information band on the top edge. They did not always give in to the band, for when the musicians produced a cover deemed to be offensive—such as the famous and extremely valuable "butcher's block" image for *Yesterday and Today*—Capitol made them change it as soon as outraged retailers began to return the record. John and Yoko Lennon's provocations met the same fate: EMI refused to release the *Two Virgins* cover on their label, although they were more than willing to manufacture the disc. *Two Virgins* was a statement of John Lennon's independence from the Beatles and their record company, and it probably suited him that so many people were outraged at the full frontal nudity on the cover.

HEROES AND VILLAINS

Capitol's *Meet the Beatles!* was drawn mainly from EMI's *With the Beatles* LP, released the previous November. Capitol had an arrangement to release EMI's recordings, but American fans did not get the same album as their British counterparts. Comparing these two artifacts gives us some insights into the operation of multinational record companies and the way they diffused popular culture. The LPs released in the United States by Capitol had fewer songs on them than the English ones, and Capitol stretched out their Beatles material as much as possible, even adding long instrumentals produced by George Martin to make up the numbers on the soundtrack albums. The album you bought in the United States was a watered-down and amended version of the original.

To Beatles fans, Capitol Records behaved very badly throughout Beatlemania—as badly as you would expect a subsidiary of a multinational company to act. Every story of a heroic rise to fame has to have a villain, and there was always a bad guy in the teen films that formed the celluloid narrative of rock'n'roll: the crooked manager, the disinterested promoter, and the deceitful record company executive. In the Beatles story, the pre–Yoko Ono villain has to be Dave Dexter, chief of A&R for Capitol records. He was the man responsible for selecting songs to be recorded, the way the various sounds of the recording were mixed together, and which songs were to appear on the LPs and in what order. Dexter has been charged with making "weird and inexcusable judgments about song choices and sequences" on the albums.[21] Certainly his compilations are nowhere as good as George Martin's, and he failed to see the importance of the concept album, but it was his company's policy to amend foreign LPs for American release. The differing number of tracks was a result of the way royalties were computed. In Europe the company paid a fixed percentage of the price of the disc, but in America, royalties were based on the number of tracks on the album. A record with fourteen tracks would cost 28 cents per track, while one with eleven tracks cost only 22 cents.[22] This gave a powerful financial incentive to reduce the number of tracks on American LPs.

Record companies in different countries send each other discs to offer the rights to release a recording, but once a deal is struck, they send the master tapes that have to be remastered by the purchasing company before it stamps out millions of copies. The mastering process allows the engineer to change sound frequencies and volume levels of the original master tapes as he transfers them to vinyl, and this can create a different sound, especially to the trained ears of engineers and musicians.

Even untrained ears can hear the differences between the Parlophone and the Capitol releases of the Beatles' records. Dave Dexter changed the original master tapes by altering the sound frequencies (by remixing the source tapes and altering the balances in the process of remastering them) and adding reverberation, an echo sound. In this way listeners on each side of the Atlantic heard slightly different versions of the Beatles' early recordings. The critic Dave Marsh is rather hard on Dexter, accusing him of "butchering" the Beatles' records, an opinion shared by other fans and audiophiles. But those clumsily remixed records, with their fake "duophonic" stereo effect and reverb-heavy sound, were the ones that captured the hearts of Beatles fans outside the UK. One of them acknowledged the failings of these Capitol mixes but added that these were the records that "we all grew up with! For good or bad, this is what most of us remember."[23]

In the eyes of fans and historians, the greatest crime committed by Dexter was not appreciating the Beatles—this was the man who turned down "Please Please Me" (which he called a "dog") and "She Loves You." Dexter is usually portrayed as a jazz enthusiast, the sort of snobby elitist despised by the Beatles and their fans, and a man who publicly hated rock'n'roll. Although Capitol was an affiliate of EMI, it turned down the opportunity to release the first three Beatles singles, to the eternal embarrassment of their management and the condemnation of generations of fans. Like the hapless Dick Rowe of Decca Records—the man who had a chance to sign the band in 1962 but told them that guitar bands were on the way out—Dave Dexter of Capitol could not see the potential of the greatest band of all time. How could anybody have been so stupid?

The story has been retold many times, and its appeal works in different ways. First, it stands for the ignorance and arrogance of top management (something we all know about), and second, it articulates the big company versus small company dialectic that forms the spine of the history of popular music. The large, integrated companies that dominated the record industry were called the "majors." They had large catalogs but mainly kept to middle-of-the-road acts with the largest potential market. On the other hand, the independents were much smaller, often amateur, outfits that met the needs of small niche markets, such as the black ghettos of American cities or the rural audience for old-time music. The independents like Sun Records in Memphis or Chess in Chicago figure large in the history of rock'n'roll. The majors were too stiff and conservative to see a new trend in pop music, so the risk was taken by an independent label in another chapter in the history of the generation gap.

Finally there has to be an element of failure and discouragement in the story of a heroic journey. After Decca and Capitol rejected the Beatles, they overcame the disappointment and persevered until the great mass of music fans realized their greatness and made them stars. History depends on hindsight, and we engage the story of the Beatles already knowing they are a great band that made timeless music. Unfortunately Dick Rowe and Dave Dexter did not have this hindsight, for they approached the Beatles with their sensibilities sharpened by the realities of the pop music business at that moment. Rowe was right: guitar bands like the Shadows were slipping in favor at that time. We have the benefit of hearing the Beatles' audition at Decca's London studios, for these tapes have survived. The miracle of recorded sound enables us to enter into the story at this one critical moment, and the songs the band performed for Decca were mediocre standards. Their playing was hesitant, and their vocals generally flat. Later that year, they auditioned for Parlophone and again failed to impress with their music.[24]

EMI did not send the Beatles' recordings to Capitol with any fanfare or recommendation; the latest Beatles release arrived in a box with about thirty others. As Charles Tillinghurst of the Law Department of Capitol recalled, they came with no special mention, and one of the tracks was matched with a song by Frank Ifield, an old-fashioned show business crooner.[25] Dexter wasn't impressed with "Love Me Do" and found nothing that distinguished it from the hundreds of pop quartets he had already heard. Capitol had fared poorly with the other English hits they had released, and this led them to believe in the current show business wisdom that English acts did not appeal to American audiences. Judging by the poor American sales of the singles that Capitol turned down, Dave Dexter was right not to release them.

The Beatles story in America begins with Vee Jay Records and their release of

"Please Please Me"—the first Beatles record available in the United States. This is the point in the story when the plucky independent comes in and saves the day by seeing genius in the music. The two companies that released the early Beatles singles were both small independents. Vee Jay Records was formed in 1953 by married couple, Vivian Carter and James Bracken, who ran a record store. They appointed record company executive Ewart Abner to run the business. Vee Jay concentrated on gospel and R&B and produced some recordings that inspired many English musicians. It was one of the few labels owned by African Americans and the first to be a major success, moving from the "race" music niche to a full line of popular music under Abner's leadership. Started on a borrowed $500, Vee Jay personified the typical independent record company: a benevolent shoestring operation mining for gold and standing up for the little guy. It got its English masters from Decca until 1962, but then EMI came in and asked if they were interested in Frank Ifield, who also recorded on the Parlophone label. The Beatles came with the Ifield deal, and Vee Jay's promotional material shows why they signed the Beatles—not because they were an exciting new band that played a lot of African American music but because they were Number 13 in the English charts. Frank Ifield was at Number 2. He got top billing and the most attention from Vee Jay, even though Ifield's nondescript crooning cannot compare today with the Beatles' recordings.

The history of Vee Jay Records is also the story the corruption and sharp dealings of the record industry, which were especially vicious on the bottom rungs of the ladder. Vee Jay did not pay the money it owed EMI for its masters, so EMI went to court to terminate the agreement. In fact Vee Jay wasn't paying any of its licensing fees or artists' royalties, nor was it paying its tax obligations or debts to manufacturers. The company was releasing hits, and the money was coming in, but unfortunately it was going out even faster. Like all the other independents, Vee Jay was bribing deejays and radio stations to play their records in the common practice of "payola." A lot of the profits that remained were being siphoned out by Abner, reputedly to pay his gambling debts.

EMI approached all the major record companies in America, including RCA-Victor, Columbia, Decca, and Atlantic, in their efforts to sell the Beatles' masters: all turned them down. EMI was scraping the bottom of the barrel when it finally made the deal with Vee Jay, who sold only 5,650 units of "Please Please Me" in the first months of its release and did not do much better with its other Beatles records; nor did the independent Swan company of Philadelphia, who released "She Loves You." Dick Clark played the single on *American Bandstand*, yet it failed to make an impression on the charts. If the story ends here, Dave

Dexter was proved right. He also turned down other Mersey Beat bands in 1963, including Billy J. Kramer and the Dakotas. "Do You Want to Know a Secret" was a Lennon-McCartney song; the record, produced by George Martin, had been a Number 2 hit for Kramer in England. Liberty Records picked it up, but this independent label could not make it a hit—more evidence that English bands did not sell well in the United States.

Capitol finally picked up "I Want to Hold Your Hand" from EMI, although many versions of the story stress their reluctance to do so. The rest, as they say, is history. Within the corridors of Capitol, several people claimed to be the one who saw the light and advocated releasing the record, including Lloyd Dunn of merchandising and sales, Brown Meggs of the New York office and even the president, Alan Livingston, who reputedly said: "I thought this group would change the whole music business!"[26] Just like at Abbey Road studios, where many engineers claimed they were the first to hear that special something in the Beatles' recordings, several Americans have also been credited with "discovering" the Beatles, including a representative of Vee Jay Records and even the performer Little Richard.[27] There are also several different versions of the process by which EMI put the pressure on Capitol to release the record. Perhaps the chairman of EMI, Sir Joseph Lockwood, personally intervened on a transatlantic telephone call to Alan Livingston of Capitol. Could it have been George Martin? Or Brian Epstein? Did Len Wood, the managing director of EMI, actually fly to New York and tell Alan Livingston he *had* to take the record?[28] Someone has to take the credit for forcing Capitol to see sense. On the other hand, the people who ran the record label might have been doing their jobs and noticed that the Beatles already had two Number 1 hits in the UK and that this might be a good time to release a Beatles record.

Failing to accept "I Want to Hold Your Hand" would have been a criminal mistake. It was that good a pop song, and although it was not made specifically for the American market, it was clear from the start that it could be Number 1 in the United States. The engineers who heard it being recorded at Abbey Road knew it would be a hit. They had already recognized the freshness and excitement of the Beatles music, which all seemed to come together in "I Want to Hold Your Hand." Geoff Emerick commented, "It was a stroke of good fortune that the group was at, or close to, their initial artistic peak when America first heard them."[29] The Beatles' timing was always perfect.

Capitol's promotional plan for the Beatles in the United States was built around their previous success, their looks, and the frenzy of the fans rather than the merits of the music. Beatlemania was portrayed as a sort of disease, and the symptoms were often expressed in terms of record sales: "Britain's 'Beatlemania' has spread to America! . . . Among record buyers 'Beatlemania' has proved absolutely contagious. Over 3 million discs already sold in England alone."[30] The look of the band, especially their haircuts, was a major element in the marketing. The "mop top" had become the Beatles' branding image in England, where the tabloids had already started to use outlines of their hair (which was called the fringe) without any facial features to represent the band. This visual shorthand was adopted by Capitol records in its promotional material.

The marketing of "I Want to Hold Your Hand" began on December 4, 1963, with a press release announcing the arrival of a hit single. For a company that allegedly did not appreciate the Beatles, Capitol certainly put everything they had into the promotional campaign, with reports varying from $40,000 (allegedly demanded by Brian Epstein before he would "release" the rights of the single to Capitol—something he had absolutely no power to do) all the way to $100,000.[31] The sum was probably close to $50,000, but whatever number one chooses, this was clearly an unprecedented amount—more money was spent on promoting this disc than on any other single in the company's history.

Capitol has never received the credit it deserved for starting up Beatlemania in America. They sent out millions of flyers (or "teaser stickers") that said "The Beatles are Coming!" and told their employees to "put them up anywhere and everywhere." It was their idea to build the image around the mop tops. They sent out plastic buttons ("Be a Beatle Booster!") and wigs in great quantities, which sales and promotion staff had to wear during the business day. The merchandising plan they developed reveals the close ties between record company, radio deejays, merchandisers, and fan clubs in promoting Beatlemania. The wigs and buttons were also distributed to radio stations and record stores to accelerate "the Beatle hair-do craze" that was reputedly sweeping the country.[32] This merchandise was added to the records and photographs given away in contests organized by radio stations and fan clubs. High school students were enlisted in spreading the word and putting up the flyers. Capitol kept up an aggressive marketing campaign throughout the Beatles' American tours, publicizing the shows in conjunction with the radio stations that presented them, and linking their promotion to television appearances. Capitol Records helped introduce the word Beatlemania into the American vernacular and worked tirelessly to exploit it.

THE RECORD AS VALIDATION

The Beatles conquest of America started with a hit record. They were performing in Paris when the good news arrived that "I Want to Hold Your Hand" was at
the top of the American charts. Their roadie Mal Evans expressed the feelings of
the band: "They felt that this was the biggest thing that could have happened."[33]
Paul McCartney has retold the story around the assertion that the Beatles would
not agree to an American tour until they had a Number 1 record in America.[34] By
the time the great news came to their hotel in Paris on January 17, 1964, all the
travel and booking arrangements were made, and they were committed to the
tour. Why create the story of such an absurd condition when the timing of the
tour made it clearly untrue? The band thought that recordings were so important
that the validation of a hit record was an essential prerequisite to achieving their
dreams in the United States.

Such was the significance of American acceptance that any British entry into
the American Top 100 was seized on by the British musical press. The strong
presence of American records in their charts was being challenged in 1960 by
homegrown stars like Cliff Richard and Adam Faith, but there were still enough
American hits to maintain the grumbling about transatlantic dominance of British cultural life. On the rare occasions that a British record went to Number 1 on
the American charts (such as Laurie London's "He's Got the Whole World in His
Hands" in 1958), it was front-page news. In the early 1960s Cliff Richard and the
Shadows was the biggest pop act in the UK, considered by patriotic music lovers
to be the vanguard of a British rock'n'roll that was just as good as the American.
But the litmus test of an English Elvis was the American response: this was the
final, definitive assessment of how close they were to the real thing, and this is
what made American acceptance so critical to English rock musicians. In 1959
ABC-Paramount issued Richard's hit "Living Doll" stateside, but it reached only
Number 30 on the *Billboard* chart. When Cliff and the band went to the United
States in 1960, expectations were high. The played on *Ed Sullivan* and toured the
country but got no response. Their records managed to enter the *Billboard* Top
100 but never broke the Top 25. Even Cliff Richard's films did not attract any attention. Their American invasion ended in a resounding defeat.

The poor showing of Cliff Richard and the Shadows was greeted with gloom
in British entertainment circles and weighed heavily on the Beatles and their
manager as they considered an American tour. We will probably never know what
attracted Brian Epstein to the Beatles, whether it was their music, their humor,
or John Lennon's rugged good looks, but we do know Epstein's goal: to make

them the "biggest theatrical attraction" in the world. To achieve this, they had to conquer America, the biggest market for popular music. If we take Epstein's autobiography, *A Cellar Full of Noise*, at face value (a dangerous assumption), we learn that he had his sights set on the American market ("the heart and soul of popular music") from the very beginning.

As the Beatles rose to new heights in their annus mirabilis (wonderful year) of 1963, the American tour assumed greater importance. It became the next logical step in their careers—not just desirable but necessary. Once the band had scaled all the heights of British entertainment—the London Palladium, the string of hit records, the Royal Variety Performance—there was only America left, the last and most difficult market to conquer, the challenge that had beaten the best of England's rock stars. The American tour became the necessary validation of their talents, not just as a theatrical attraction, but as rock'n'rollers, as practitioners of the American music they had admired for so long. To take it back to its homeland and receive the acclamation from the people they most admired really appealed to them. Brian Epstein recognized the "turning point" as the moment they touched down at Kennedy Airport: "We knew that America would make us or break us as world stars . . . In fact, she made us."[35]

CHAPTER 2

BEATLEMANIA

As Ringo Starr pointed out during the Beatles' first tour of the United States, the Americans had all gone out of their minds. This sentiment was repeated by his band mates and everyone else there, from the promoters to the security detail. The cop who told an English reporter "I think the world has gone mad" summed up the general reaction to Beatlemania.[1] The mass hysteria when the group first touched down at the newly named Kennedy Airport in New York City on February 7, 1964, was only the beginning of the highly visible public events that would dwarf the adulation previously directed at pop stars. Young women had screamed and hurled themselves at Elvis Presley and Johnny Ray in the 1950s; they had idolized the young Frank Sinatra in the 1940s; and they had even danced in the aisles to the swing music of Benny Goodman in the 1930s, but none came close to the scale and ferocity of the public outpourings of affection directed at the Beatles in 1964 and 1965.

The band was a national obsession in the United Kingdom, yet no one could have imagined the response from the United States; it even caught the Beatles and their management by surprise. In addition to being seen as "one of the most extraordinary and significant events in the history of American show business," Beatlemania marked a turning point in millions of young lives. It was a magical time, a historic moment in the minds of an influential generation, and people who were there tell us that you had to be there to experience it. For those who were there, the Beatles were changing the world, and as one fan said later, "There is no question in my mind that The Beatles had the MOST profound impact on history."[2] Since those halcyon days, nobody in popular entertainment has been able to repeat this moment in all its economic and cultural significance.

Beatlemania was bigger than the Beatles, drowning out the music and eventually overshadowing the musicians themselves. It was much more than screaming fans. Entertainment industries capitalized on it to promote concerts, records, films, magazines, books, clothes, and toys. It inspired countless copy cat bands, ranging from the very successful Monkees to the quickly forgotten Liverpools and their *Beatlemania in the U.S.A.* record. Beatlemania was one of the largest and most successful merchandising campaigns in American history. As one magazine explained, "Today's modern Beatle fan can wear Beatlemania, speak Beatlemania, play Beatlemania and even eat Beatlemania."[3]

In the long run Beatlemania was a significant chapter in the history of celebrity, a major cultural event that highlighted important social trends, like the rise of the counterculture. The Beatles' music went far beyond mere entertainment, acquiring significance and providing meanings deeper than the lyrics of those well-remembered songs. In a decade remembered by its popular music, the Beatles have become an essential part of our collective memories of the 1960s. Understanding Beatlemania is part of making sense of that tumultuous decade.

Madness. That was Beatlemania. *Time* magazine called it "The New Madness." Eyewitnesses found it difficult to describe this "frenzied scene that beggared belief," and as many observers pointed out, it was impossible to imagine it if you had not seen it with your own eyes and heard it with your own ears.[4] Larry Kane was a deejay on a Miami radio station who got the opportunity of a lifetime to join the press corps accompanying the Beatles during their first two historic tours. In a book appropriately titled *Ticket to Ride*, he described the frenzied crowd behavior that was repeated all over the country wherever the band played. He remembered screaming so loud it hurt his ears. Girls screamed at top volume for the entirety of the Beatles performance (that clocked in at around thirty minutes), maintaining a high-pitched cacophony from the moment Ringo Starr's drums were brought onto the stage until the final desperate news that the band had left the auditorium. The noise levels were painful and often compared to thunder or an earthquake—the waves of sound an oppressive, physical force, beating down on people, forcing their hands over their ears. A reporter described it as a jet engine streaking through a summer thunderstorm: "It had no mercy."[5] The screaming drowned out all in its path, and everyone left the show with a ringing in their ears that lasted for a day or two.

Larry Kane was shaken by the almost animalistic behavior of the fans—nice young affluent Americans with good schooling and careful upbringing. He was alarmed by the tidal waves of young female bodies hurled against the stage: "Girls and some boys close together, standing screaming, moaning, groaning, ripping at their hair, pushing, shoving, falling on the floor and crying, real tears streaming down hundreds of faces, smearing their mascara and lipstick, and mothers and fathers hiding in the back, some of them dancing to the music."[6] The young women were usually described as hysterical or frenzied—squealing,

The Beatles arrive at Kennedy Airport to begin an American tour in 1964. Behind them stands their manager, Brian Epstein; in front of them are thousands of screaming girls. Paul and Ringo are looking upward at the packed galleries of fans that had spent hours waiting for the arrival. (New York Daily News Archive, Getty image 97269347, courtesy Getty Images)

wailing, screaming, weeping, and beating their heads with their fists in the agony of not being close enough to their idols: "I remember ripping part of my hair out of my head, screaming—we couldn't talk after the concert we were screaming so bad."[7] Some girls laughed with joy; others cried within the "deafening ecstasy" and "screaming pandemonium" of the Beatles' audience. A considerable number were overcome by this massive assault on the senses: "I had fainted. I think the emotions, the hot August night, it was the first time I saw them, it was exciting."[8]

These were not seasoned rock fans; many of them were attending their first concert. Although the 1960s are remembered as a decade of permissiveness, the fans' personal accounts of Beatlemania reveal that much of the strict 1950s was still in place in 1964. For many of the lucky ones who went, not only was it the

Caught in mid-shriek, these two girls were among the 55,000 fans who packed into Shea Stadium in August 1965 to watch a historic Beatles concert. According to the promoters, this was the biggest crowd, and the biggest gross, in the history of show business. It was also the loudest. Security guards had to cover their ears because of the intensity of the crowd noise. The Beatles were deafened by the screams and could not hear themselves play. Many of the fans are holding cameras; the girl on the right holds an 8 mm movie camera. (New York Daily News Archive, Getty image 97340351, courtesy Getty Images)

first show they attended, but it was the first time out without a chaperone, and even the first time they got to stay up after 11 p.m. Seeing the Beatles in person was an important step in growing up, the significant event that began the sixties for them. The girls came in their best clothes: "We wore canary yellow dresses so that the Beatles would see us, and [we] would stand out. And apparently everyone else had the same idea," and brought with them Beatles' merchandise, photographs of their favorite band member, cameras with a supply of flashbulbs, homemade signs, and gifts and tributes to hurl at the band on stage.[9] During the tour all four musicians were hit by flying objects, John, Paul, and George sustaining cuts and bruising while Ringo cowered behind his drums. When a teen magazine revealed that George Harrison liked jelly beans, the band was bombarded with them—about two tons a night as George remembered—while they tried to play. This practice spread by word of mouth and the press (although the Beatles went out of their way to discourage it), and it became "the obligatory throwing of jelly beans," a ritual invented by the fans that was repeated at every performance. Other objects thrown at the band included food, dolls, clothing, and used flash bulbs.[10]

Such was the emotional energy generated by the Beatles that more than one concert turned into a riot. Squads of police, some on horseback, had to hold back the crowds—no easy task when you consider the size of the attacks they had to repulse, which ranged from five hundred to several thousand (the police estimated that about six thousand fans stormed the stage one night), and the feral determination of young girls to get close to their idols. The police used dogs, horses, nets, barricades, lassoes, and their batons in attempts to control the fans. In two concerts in Cleveland, the Beatles faced full-scale riots of thousands of people and had to flee for safety. Mindful that these dedicated fourteen-year-olds "could break through the defensive line of the Cleveland Browns if they wanted to," the police had to threaten to cancel the show to restore order.[11]

At Vancouver about 135 young people were injured in a melee during the show, suffering broken limbs, concussions, contusions, and what Larry Kane described as the consequences of Beatles-induced combat—"bloody lips and noses, bruises, welts and abrasions." In the aftermath of a rock concert, one expects to find litter, food wrappers, and empty alcohol containers, but Kane noticed that the food concession stands at a Beatles concert did very little business—the audience did not want to miss a minute of the show. Instead he saw gauze, medicine bottles, and bandages left by the temporary triage stations set up at the back of the concert halls as well as spent flashbulbs and lost shoes scattered around everywhere.[12] After the Beatles' Cleveland concert at a baseball stadium, hundreds of shoes (singles not pairs) were piled on top of the pitcher's mound. Their official biographer, Philip Norman, described performances as "cops and sweat and jelly beans hailing in a dream like noise . . . faces ugly by shrieking and biting fists; it was huge amphitheatres left littered with flashbulbs and hair rollers and buttons and badges and hundreds of pairs of knickers, wringing wet."[13]

The fans' enthusiasm for the music was so intense that it produced moments of sheer terror when all that adolescent energy and excitement threatened to turn into violence. When John Lennon was asked how he felt after a near riot at the Cow Palace in San Francisco—the first date of their second American visit of 1964—he confessed: "Not safe. Can't sing when you're scared for your life." (But this was also the person who encouraged a noisy response: "I like a riot.")[14] Larry Kane thought that the Beatles seemed to be running for their lives when they left the stage. When they got to the end of their last song, they unclipped their guitars from the straps on their shoulders, and the moment the music ended, they dropped their instruments and ran.

The madness was not restricted to the concerts, and this gave Beatlemania the extra dimension that lifted it above all other incidents of fan worship: the home invasions, the besieging of hotels, and the assaults on recording studios. Fans mobbed the Beatles as they traveled to and from concerts. They permanently picketed their hotels on tour and their residences in the UK. Often they physically attacked the musicians, ripping off their clothes and forcibly removing pieces of their hair. They took anything that could be torn off, including Ringo's St Christopher's medal. As one of their entourage admitted, "To see three thousand almost deranged girls heading your way was quite terrifying." And the consequences to the band if they were caught by this mob were frightening: "They could be stripped naked and knocked unconscious . . . or worse, scalped."[15] In San Francisco their limousine did not leave the stage quickly enough, and the weight of fans on the vehicle caved in the roof. In Houston fans broke through the barricades and climbed on the wing of the Beatles' airplane before it had reached the terminal. After the mayhem of two concerts at Forest Hills Stadium, critic Robert Shelton warned that the Beatles had created "a monster" and that "they had better concern themselves with controlling their audiences before this contrived hysteria reaches uncontrollable proportions."[16]

By necessity the security arrangements for their concert tours were as meticulously planned as military operations, involving elaborate ruses like decoys and deliberately misleading the waiting press corps. Some American cities gave the Beatles the level of protection reserved for the president; others provided more. Ringo once said that being a Beatle was like being in "the bloody secret service."[17]

One journalist compared the band's transportation plans to a shipment of gold from Fort Knox. The security involved hundreds of police officers, miles of cyclone fencing, and even an iron cage to protect them en route. They traveled in police cars, ambulances, armored cars, and even fish trucks to avoid the mob. They exchanged clothes with policemen to dupe the waiting throngs and often disguised themselves if they had the nerve to leave the hotel and walk outside.

The positions of road manager and personal assistant, filled by Mal Evans and Neil Aspinall respectively, were dangerous jobs that required a mixture of careful planning, guile, and courage. After a few months of field testing, they chose limousines for ease of entry: the Austin Princess had the widest doors, and it was easier to throw people into them. The work was more than a matter of conveying the musicians and their valuable equipment; it involved defending their charges against threats as diverse as fans hidden in hotel rooms (some armed with knives), mothers of fans getting stuck in air-conditioning ducts, photographers who stormed dressing rooms, and attractive young ladies who were not going to take no for an answer. (The job was not without the perks that came with hundreds of women willing to do anything to get close to John, Paul, George, or Ringo.)

The rest of America struggled to find the words to describe Beatlemania and the reasons for such "dangerous mass hysteria among young people." The record industry likened it to a disease spreading rapidly through the teenage population, "and doctors are powerless to stop it." They spoke of a "sales epidemic" that had swept through Europe and now was appearing in North America. *Newsweek* reported that the concerts were "slightly orgiastic" because of the power of amplified electric guitars.[18] Beatlemania was interpreted as a religion by commentators trying to find the root causes for so much teenage emotion. Ken Ferguson described it as a cult and "a form of hysterical worship" in an article written for *Photoplay* magazine. He experienced the "full blast and fury" of the fanatical fans who had been mesmerized by the Beatles' "savage, pulsating, hypnotic sounds." Brian Epstein told the press that he was going to the United States to "spread the gospel" of his band.[19] Some of the Beatles' most devoted followers experienced a powerful spirituality that went beyond the music. Throughout the tours parents brought sick, crippled, and blind children backstage to receive the healing touch of the four embarrassed musicians. John Lennon told his friend Pete Brown that dealing with these sick children and their parents was by far the worst part of the tours.

The fans were the objects of both fascination and disdain by those adults who observed them. The conservative papers *Daily Express* and *Daily Telegraph*, the bastion of British middle-class fears and values, deplored the animalistic behavior of the young and voiced concerns that these masses of easily impressed youth might be open to more sinister suggestions than "She Loves You." In an editorial, the *Telegraph* compared Beatlemania to the hysteria and barbarian fantasies of the Nazi's rallies at Nuremberg. The fans were seen as mindless, "pitiable victims," hypnotized by their grotesque idols: "the huge faces, bloated with cheap confectionary and smeared with chain store makeup, the open sagging mouths and glazed eyes, the hands mindlessly drumming in time to the music." Paul Johnson described them as pathetic, dull, and vacuous—the least fortunate of their generation.

What Beatlemania represented to critics like Johnson was the triumph of the entertainment industry, a commercial machine that had turned a generation into "fodder for exploitation." These views followed the criticisms made by Theodor Adorno and Max Horkheimer of what they called "the culture industry" back in the early twentieth century. The purely capitalist operations manufactured standardized mass entertainment that not only brought profit but also provided ideological legitimization of the system. Adorno thought that mass entertainment integrated consumers into capitalism and manipulated them into a process of response mechanisms that undermined the ideal of individuality in a free society. Conformity had replaced consciousness, and this process depended on the star system propagating "great personalities" and "heart throbs."[20] Beatlemania is the outstanding example of the effectiveness of these culture industries.

CREATING BEATLEMANIA

Americans might have established Beatlemania as the significant cultural event of the decade and made the band global celebrities——but they did not invent it. Beatlemania emerged in 1963 as the Beatles rose to the very top of British entertainment in an amazingly short time, capturing the attention of nearly everybody on the island, from preteens to the royal family. During 1963 the voracious English press, especially the mass-circulation tabloids, weaved together this annus mirabilis of record sales, chaotic concerts, and screaming fans into a major news item. English newspapers did not report about popular music and only noticed rock'n'roll when it inspired incidents of delinquency. The up-market broadsheets like the *Times* kept to their reviews of classical performances, while the tabloids thought up nasty words to describe the sound of the new youth music. The press interpreted rock'n'roll as another teenage craze, "The Big Beat Craze" that the *Daily Mirror* expected to burn out quickly: "Beat Bubble Could

Burst Overnight." Yet pop music celebrity grabbed the tabloids' attention with the excesses (financial, sexual, and alcoholic) of the rock'n'roll lifestyle, which seemed as out of control as the fans. Like their counterparts in the United States, the English newspapers were always interested in the rags to riches (and then back to rags) story, and popular music was becoming one of those mythical but carefully observed domains where fame and money could transform the lives of ordinary people with the magic wand of stardom.

The rise of Elvis Presley or Cliff Richard had provided some titillating copy for the press, but this was nothing compared to the Beatles, who exploded from an obscure regional pop scene into national prominence almost overnight. As John Lennon's half-sister Julia Baird remembered about the friends and family watching from Liverpool, "We knew they were going to be big, but we didn't know what big was." This sentiment was repeated in London by the professionals at EMI who saw the Beatles break all the records for sales, concerts, awards, and fan mania.[21] The band quickly moved into other arenas, such as film and television, and repeated their triumphs. Their star had ascended so quickly and shone so brightly that it overpowered every other one in the entertainment firmament. As the band continued to reinvent what it was to be rich and famous, Beatlemania grew larger.

The story had appropriately humble beginnings. Exactly a year before they touched down in New York, the Beatles played the Regal Theatre in Wakefield—a bleak and gloomy town in the industrial North that does not even justify a mention in the British Tourist Association's *Touring Book of England*. The night before, they had played Bedford in the Midlands, and after Wakefield, they had to travel all the way up to Carlisle, which is on the border with Scotland. They were one of the supporting artists on a package tour headlined by Helen Shapiro, a sixteen-year-old who already had several hit records, an American tour (including an appearance on *The Ed Sullivan Show*), and a film to her credit. The Beatles were way down at the bottom of the bill, but they were still happy to be on a bill anywhere—at this point in their career, the object of the exercise was to keep working, and it didn't really matter where. Outside Liverpool, they were pretty much unknown. They could walk about the streets of Wakefield unmolested and grab a cup of tea and a sandwich at a greasy café before going on stage.

From a strictly historical viewpoint, the first buds of Beatlemania had broken through in the early months of 1961, when the band members realized they had made an important transition. George Harrison said, "People were following us around, coming to see us personally, not just coming to dance."[22] They had returned from an extended stay in Hamburg as a greatly improved band, and the girls at the Liverpool clubs they played were indeed taking notice of them. In an amazingly prescient comment made in 1961, a writer for the Liverpool music newspaper *Mersey Beat* called them "the stuff that screams were made of," blissfully unaware of the noise that was to follow.[23] The Beatles were big in the Liverpool music scene, but Liverpool was a down-at-heel northern town that nobody in the music business took seriously. London was the center of popular entertainment, and Liverpool, as one record company executive put it, might as well have been Greenland.

The Beatles made their first record in 1962, an achievement that greatly pleased the four musicians, and for at least two of them, this might have generated enough pride and satisfaction to justify the whole journey had it ended then. They also made their first radio broadcasts. They acquired a new manager, got rid of their first drummer, Pete Best (something that happened all the time in pop groups), and finalized their lineup around three guitarists and their new drummer, Ringo. They were voted the best band in Liverpool in a *Mersey Beat* poll and drew impressive crowds whenever they played in their hometown. They also made a triumphant return to Hamburg—a return engagement that emphasized how far they had come from the loose band of amateurs who had arrived two years earlier with little idea how to play and no idea of what to expect. They ended the year with a successful recording session at the Abbey Road studios of EMI.

In January 1963 the police had to be called in to keep order at a concert. The Beatles' second record, "Please Please Me," entered the British charts in February at Number 16, sandwiched between the latest American singles from Brenda Lee and Duane Eddy. But by the end of the month, it had reached the top position. Along with thousands of other young musicians, the Beatles had dreamed the dream of reaching the top of the charts, but they had no idea how much it would change their lives. The English music press began to report that the fans were going wild. The band had started at the bottom of the bill on the Helen Shapiro tour, but now the kids were calling out "We want the Beatles!" throughout the shows. Articles about them and their fans began to appear in the press. In February Maureen Cleave of the London *Evening Standard* wrote a story that tried to answer the question on everyone's mind: "Why the Beatles Create All That Frenzy." As "From Me to You" raced up the charts, the screams increased. The Beatles hardly noticed that their anonymity was slipping away. By the summer of 1963, thousands of fans were turning up at the stage doors of the small venues they played, blocking the entrances and threatening to riot.

In October they appeared on the country's most popular televised variety show, *Sunday Night at the London Palladium*—the British version of *The Ed Sul-*

livan Show. The whole country watched them play four songs, including their latest release, "She Loves You," but far more important to the newspapers was the crowd outside the theater. The hundreds of young girls who appeared at the stage door of the Palladium well before the concert had swelled to an unruly mob by the time the Beatles turned up. The next morning, October 14, the story dominated the English press. The *Daily Mail,* the *Daily Herald,* and the *Daily Mirror* made a great deal out of "the fantastic Palladium TV siege," and undoubtedly exaggerating the number of "frenzied" fans, but nevertheless the pictures of them were dramatic.[24] Even the stuffy *Daily Express* carried the photographs.

On October 21, the *Daily Mail* put the Beatles on the front page and on page three used the term "Beatlemania" for the first time. By November it was in general use, and the band and its fans were major news items. The stories focused on the "squealing females," the "hordes of kids" that materialized as soon as the Beatles were spotted, followed by the rugby-like scrums of bodies, with the policemen's helmets rolling on the pavement, and the frantic escape of the musicians to cars that were almost submerged by mobs of young people.[25] As newspapers warmed to Beatlemania, they regularly reported the number of fans in attendance and usually inflated them, with estimates running as high as ten thousand. Even the staid *Sunday Times* got into the act, describing five thousand "hysterical youngsters" who had slept out on the pavement days before the Beatles arrived. With only an occasional cynical aside that these demonstrations might have been staged, the press eagerly joined in the enthusiasm for Beatlemania: "You have to be a real sour square not to love the nutty, noisy, happy, handsome Beatles!" said the *Daily Mirror,* doubtless trying to connect with a younger readership.[26]

The story in the English press gradually got some attention in the United States—a country that in 1963 was concerned with far graver issues than a pop group from Liverpool. So the first signs of Beatlemania were small. Martin Goldsmith provides an appropriate metaphor for its dawning: "Like an approaching thunderstorm, the Beatles' arrival in America was preceded by a few low rumbles and flashes of light, none of them giving more than the slightest hint of the potency to follow."[27] Articles in the press and on television often described the effect of the Beatles in terms of noise and light, as extraordinary occurrences in the natural world, and as signs that came as a premonition of something important to come.

Beatlemania began to seep into the American press by the end of 1963. On November 15, *Time* magazine ran a short story on the excesses of the fans. This was followed up by *Newsweek* on the 18th, which described their music as "one of the most persistent noises heard over England since the air raid sirens [from

World War II] were dismantled." Then *Life* ran several articles in December and January, which leaned heavily on images taken of the fans in England. One eight-page article in *Life* carried seventeen photographs, and most were pictures of the fans and the police cordons they assaulted.

All three American television networks became interested, and *CBS Evening News* took the lead by broadcasting a three-minute story mainly about the "adolescent adulation" of Beatlemania that was the "modern manifestation of compulsive tribal singing and dancing." CBS repeated this story later in December, when the grief that followed the assassination of President Kennedy had abated slightly. By the time the press and record companies were preparing for the Beatles' arrival, Jack Paar (one of Ed Sullivan's competitors) ran some short performance clips on his show on January 4. He made light-hearted fun of the fans and said later that he had no idea that the Beatles "were going to change the culture of the country with music . . . I brought them here as a joke."[28] Americans were learning about the Beatles not from listening to their music, but from the news reports about their fans.

What got Beatlemania noticed was its scale. When Pan Am Flight 101 left London airport, about one thousand fans screamed their farewells, but an estimated three to five thousand welcomed them to New York. The Beatles were somewhat apprehensive about their reception and thought that the throngs at the airport were there to meet someone else, a head of state or somebody like that. A harried airport official commented, "We've never seen anything like this before. Never. Not even for Kings and Queens."[29] This response, framed in shock and awe, was repeated throughout the tour. It came from security men, hotel staff, theater managers, and transportation officials: "It scares you . . . It's just beyond me. I've never seen anything like this," "I've never seen anything like it, and I was here when Castro arrived, when Khrushchev came in, but this topped them all," and "This is kind of like Sinatra multiplied by 50 or 100."[30]

Despite the fervor of the American welcome, and the avid attention of the media, the people who were most excited on that first day in New York City were the Beatles themselves. They might have been surprised by their fantastic rise to fame in the UK, but these accomplishments paled against their amazement and delight at their American reception. On the journey between Kennedy Airport and the Plaza Hotel, Paul McCartney experienced feelings of elation: "It was like a dream. The greatest fantasy ever."[31]

A TELEVISED SPECTACULAR

When the Beatles finally established their tour headquarters in New York, one hundred thousand fan letters awaited them—all written by people who had never heard them perform a live concert. Ed Sullivan already had 50,000 applications for the 728 seats in the venue where his show was broadcast. He was not sure what he was getting when he signed the Beatles for three performances. Like other impresarios, he regularly crossed the Atlantic to look for talent for his variety show, and when he saw the Beatles fans going wild at Heathrow Airport, he told his entourage that this was exactly the same excitement that Elvis had aroused. By the time the broadcast was ready to go on air, Sullivan had overcome his second thoughts about top billing for an unknown band and was beginning to grasp the significance of the event he was about to present to the American people. Beatlemania had made its mark on America in just a few days. Along with all those other grizzled professionals who had spent their careers in the capital of the entertainment industry, Sullivan was amazed at the frenzy of the fans. He told his viewers, "All these veterans agree with me that the city never has witnessed the excitement stirred by these youngsters from Liverpool."[32]

The screams began the moment he introduced them and continued through the show. After they finished their last song, "She Loves You," a bemused-looking Sullivan read out a congratulatory telegram from Elvis Presley (which had been composed by his manager, Colonel Tom Parker, without his knowledge). Presley's performances on Ed Sullivan had been a milestone in his career and in the acceptance of rock'n'roll by mainstream America, and this telegram was the symbolic passing of the baton to a new generation. In addition to Elvis's famous appearances (when he was shown only from the waist up), Buddy Holly and Bill Haley also enjoyed historic appearances on the Sullivan show—the latter's performance of "Rock Around the Clock" in August 1955 was the first time millions of adult Americans experienced the excitement of rock'n'roll. But nothing compared to the impact of the Beatles. The 73.9 million people who watched on that frigid February night were the largest television audience ever recorded up to that point, far exceeding the 60 million people who tuned in to Elvis. The Beatles had captured the attention of over a third of the population of the United States and a very high proportion of the country's 22 million teenagers. In New York City, three-quarters of the city's television sets were tuned into Ed Sullivan, and it was said that even the criminals took time off to watch the Beatles.

All over the country, there were millions of individual epiphanies as Beatlemania took hold: "When they struck that first chord it just sent something through

The Beatles in Washington, D.C., in February 1964, as captured on the cover of one of the many magazines and books that appeared during their various American tours. These mass publications hastily combined photographs and text and often included "personal" messages from the Beatles to give the purchaser a feeling of closeness with them. The signed message in the magazine *Beatles in America* said, "We made friends with a whole country of strangers by television." Quite true.

me," "I was two inches from the screen, screaming," "I sat with my girlfriends to watch them. We were feeling the TV and touching it and screaming." The magic was not restricted to teenage girls, for four-year-old Jay Willoughby "went berserk!" with excitement. He more or less decided right then what he wanted to do with his life.[33] So did hundreds of thousands of other kids, sitting trans-fixed before tiny black-and-white screens. How many careers in pop music were activated that Sunday night we shall never know, but several important bands formed as a result of watching the program, including the Byrds and Creedence Clearwater Revival. A young man who watched the show from the wings went on (as Davy Jones) to become part of the Monkees. There must be a legion of rock stars and amateur strummers out there who wistfully look back to the Beatles on Ed Sullivan as their starting point. None of them had any idea that you could be a rock star until they watched the Beatles that night.

As important to Beatlemania as this broadcast was, the next day, a school day, millions of kids realized that something important had happened during this shared media moment: "The next day at school, that's all anybody talked about. And all of a sudden all of the boys that had their hair slicked back on Friday—on Monday it was all combed down."[34] Monday marked the formation of a nation-wide Beatle-lovers community, as young people were drawn together by their shared experience. Seeing the Beatles for the first time was described by them as a religious conversion that brought feelings of elation and joy. In the months to come, Beatlemania evolved into a youth culture, where "Beatlepeople" interacted in "Beatleland."

The superlatives that followed the Ed Sullivan broadcast increased with ev-ery new triumph in 1964, and everywhere the Beatles went, the crowds were bigger, the screams louder, and the pandemonium harder to comprehend. An American promoter, Sid Bernstein, had convinced Brian Epstein to put on two shows at the historic Carnegie Hall in New York—one of the most prestigious concert halls in the world. Part of the myth of Beatlemania was that no rock act had ever been allowed on its stage, not even Elvis Presley, but the careful research of Bruce Spizer shows that to be incorrect; several rock'n'rollers, including Bill Haley, played there in 1955. But no matter, the "first" at Carnegie Hall was added to the list of unprecedented achievements of the Beatles.[35]

Bernstein took an enormous risk in signing a virtually unknown band (at the time he made the deal over the phone with Epstein in October 1963) to a very large hall for two consecutive concerts. How he convinced the Carnegie Hall management to accommodate this group of long hairs from England is unclear, but he made no mention of rampaging fans. By the time the concerts were played in February 1964, *The Ed Sullivan Show* had made Bernstein's gamble a sure thing, and afterward he upped the ante by contracting with the 55,000-seat Shea Stadium for another concert in August 1965. Despite the unimaginable success of the Beatles up to this point, Brian Epstein was quite understandably concerned that no act in show business could fill a huge stadium, but Bernstein offered him $10 for every unfilled seat, and the deal was made. The show was a resounding success and a milestone in the entertainment business—the pop concert had now become an event. Bernstein "saw the top of the mountain" at Shea and so did scores of other promoters who realized that the sky was now the limit in youth music. Brian Epstein told the British press that the triumph of the Beatles was so total, so complete, that it would never be matched again. In his words, the Beatles had made history.[36]

The *Daily Mirror* called the Beatles' American tour "The Most Astounding Triumph in Pop History."[37] Superlatives like these were the defining feature of Beatlemania that elevated it above all other outbreaks of fan worship. At first the Beatles, like every other band, were compared to Elvis Presley—the gold standard in record sales and fan hysteria. They were introduced as the biggest sensation in pop music since Presley, but as they kept on breaking all the records that defined success in the entertainment industry, they were anointed as the biggest thing ever, "the most successful rock and roll band the world has ever seen." Since then the Beatles' yardstick has been used to evaluate all other pop acts.[38] The Shea Stadium concert had a special importance, and everyone there saw the significance of that night: "That's when it hit me. This is a big deal. This is no little rock 'n' roll band."[39]

No one had taken Brian Epstein seriously in 1962 when he patiently explained to all who would listen that he sincerely believed that the band "will be the biggest thing entertainment has ever known." Now journalists, promoters, and broadcasters were saying the same thing, only louder. Beatlemania took on a life of its own as it fed off the self-awareness and aspirations of its multitude of followers: "The Beatles . . . had the lyrics to change the world. Beatlemania! It was and is very real."[40] The people around the Beatles sensed something different about the band. The musician and critic George Melly had the thankless task of introducing them on the Helen Shapiro tour. As a Liverpudlian and trad jazz player, he was in a better position than most to appreciate their appeal, and he was struck by something special in the Beatles.[41] Their American fans had no doubt: "At the airport, you could really feel that something was happening. We had been bored, and you felt like this was the beginning of the earthquake." Some of the girls were convinced that they were part of something important: "I know it's gonna go big and it's gonna go far."[42]

"Revolutionary" was soon being used to describe the Beatles. First they had revolutionized the production of music, then the world of entertainment as they brought more acceptance of youth music and more legitimacy for young musicians. Some even saw a social revolution in Beatlemania.[43] The Beatles and their management did make significant changes in the business of entertainment, pioneering the one-act performance, the music video, the stadium show, and the world tour. No musicians in the 1960s did a better job of merging music with image, and no entertainer could command as many star vehicles as the Beatles. By the end of the decade, they had enlarged the status of the pop star and helped make rock'n'roll mainstream entertainment. The bands that followed the Beatles enjoyed an entirely different environment and much greater ambition. The paradigm of success created by Beatlemania would dominate popular music for the next two decades.

EXPLAINING BEATLEMANIA

As much as the attainments of the band were seen as unprecedented, it was the behavior of the fans that really shocked observers and parents. As these outsiders recoiled from the outpouring of assertive sexuality from tens of thousands of young females, they reached for words like "revolutionary" to explain it.[44] An army of social scientists rushed in to make sense of Beatlemania. In an article in the *New York Times*, John Osmundsen broke down "every standard explanation in the book" from psychologists, sociologists, and anthropologists. The most common explanation he found was that teenagers needed to have a good time as a relief from the anxieties of living "in an uncertain world plagued with mortal dangers." Beatlemania took a generation's mind off the threats of the Cold War and acted as a relief valve for suppressed tensions and anxieties.

There also appeared to be links between Beatlemania and the growing affluence of teenagers and the rise of a consumer society. Yet socioeconomic theories were usually trumped by more emotional explanations. Anthropologists and journalists liked to depict Beatlemania as a tribal rite accompanied by jungle rhythms, a throwback to the "uninhibited, kinetic self expression" of primitive man. The musicians played the role of witch doctors, "who put their spell on hundreds of shuffling and stamping natives."[45] Adolescent psychology was a fertile and quite profitable ground for social scientists. Pop psychologists like Dr. Joyce

Brothers, who plied their trade in newspapers and television, saw Beatlemania as another rebellion of youth against their elders. This was a popular theme of the 1950s, related to concerns about changing social roles and fears of delinquency.

Beatlemania did represent revolt to many Americans, mainly parents, but strong patterns of conformity in the fans' behavior fit the analysis of Theodor Adorno. In his theories, the fans' obedience to the beat expressed their desire to obey. There was a measure of conformity in not conforming! Sociologist Renee Claire Fox of Barnard College saw the contradictions of American society reflected in Beatlemania. The Beatles' androgynous looks contrasted female with male characteristics, and they managed to straddle both adult and juvenile worlds. They were good kids who posed as bad boys.[46] The last word has to go to John Lennon, who refused to explain the phenomenon, preferring "to leave it to the psychologists and let them get it wrong."[47]

Since the 1960s numerous accounts of Beatlemania have attempted to explain this phenomenon, ranging from scholarly books and articles written by professional historians and sociologists all the way to impressionist accounts produced by fans and posted on the Internet. Two main explanations of Beatlemania have emerged from these accounts. The first attributes the success of the Beatles as a relief for the depression that followed the assassination of President Kennedy. Thus the Beatles came to "a wounded country in its time of trouble" and "helped dispel the gloom of that death in November."[48] The second explanation places the success of the Beatles within the context of the stale and empty pop formulas that characterized American youth music at the time. In the words of Jonathan Gould, these were the "Dark Ages of Rock," the age of prepackaged teen idols, lush orchestral accompaniment, and juvenile novelty songs.[49]

All these explanations have firm foundations in the reminiscences of the people who were caught up in Beatlemania. The great national sadness that followed the death of John Kennedy was felt especially hard by teenagers. A youthful and optimistic president had spoken directly to young Americans, giving them hope and pride in this "new generation" that was making important changes in the world. Some of that hope and empowerment died with the president. After more televised bloodshed came the painful realization that the United States was a hopelessly violent and hateful place. And then came the Beatles, in the words of musician Jerry Garcia, "a happy flash. Post Kennedy assassination. Like the first good news."[50] Richard Manley remembered that the two months after the Kennedy assassination were "a sad scary time for thirteen-year-olds like me." Ron Monteleone thought that "the whole nation was in a deep depression over it. The Beatles came along and lifted the world's spirits."[51] For music critic Lester Bangs, they were the "perfect medicine" and "a welcome frenzy to obliterate the grief with a tidal wave of Fun [sic] for its own sake which ultimately was to translate into a whole new hedonistic dialectic."[52]

The gloom after the death of President Kennedy was largely felt in the United States, while Beatlemania was a global phenomenon, equally powerful and pervasive in places like Australia, which brought out the largest recorded gatherings of fans. Although the sadness after the tragic death of the president might have prepared the way for the joyous reception of the Beatles, it seems rather a stretch to assign the causes of Beatlemania to the aftermath of an assassination, especially when more than two months had elapsed before the Beatles arrived. This was the argument used by Ian Inglis in his careful debunking of this theory. He listed all the other significant events in the six months that followed the assassination and asked if they could be explained as a consequence of this event.[53] It cannot be disputed that many of the fans who took part in Beatlemania looked back at the joy as an antidote to the unhappy times, but plenty of others don't agree with the connection. There can be no better representative of the fans than author Bruce Spizer. He remembers his sadness after the events of November but adds, "by the holiday season of 1963, I was over it."[54]

As for the paucity of quality music available to American audiences in the early 1960s, many people scathingly affirmed that the Beatles had landed in a country with very bad pop music: "I was waiting for something. Nothing was going on. It was crappy music like Paul Anka." Then the Beatles' performance on *The Ed Sullivan Show* opened eyes and ears to something much better: "I was 13 and in love with Fabian and no band from Britain was going to change that until Mom put on CBS."[55] Performers like Paul Anka, Frankie Avalon, Fabian, and Bobby Rydell produced the bland, safe popular songs that had replaced the energy and raw excitement of rock'n'roll. A few years later Don McLean sang about "the day the music died" in his song "American Pie." The passing of Buddy Holly was indeed a great blow to rock'n'roll, and after him came a deluge of mediocrity. If the Singing Nun could get to Number 1 in the charts with "Dominique" in November 1963, surely the time was right for a return to the excitement of rock'n'roll?

The Beatles came to the United States with some fresh sounds and finely crafted records, but was the competition that bad? American popular music of the early 1960s contained some outstanding creativity. These were the golden years for Motown and for artists like Ray Charles and James Brown. Even though Don McLean and his audience could lament that the music had died, more than enough rock'n'rollers were keeping the beat of Elvis and Buddy, such as Tommy Roe ("Sheila") and Del Shannon ("Runaway"). There were plenty of slicked-back crooners and fabricated pop records like "The Twist," but among them were some well-written songs that have stood the test of time: "Blue Velvet" (Bobby Vinton) and "Breaking Up Is Hard to Do" (Neil Sedaka).

The fans waiting for the Beatles at Kennedy Airport were ready for something different. It was in the air, a shared expectation that bound together thousands: "It was very electric, it really was, like something was going to happen."[56] Perhaps the key to explaining Beatlemania is timing, being in the right place at exactly the right time to exploit numerous strands of development that were coming together to form a critical mass. To unlock this particular puzzle, we have to go back before the 1960s and examine the business and technological networks that brought the Beatles to America and primed their audience for the experience of a lifetime.

CHAPTER 3

LIVERPOOL

John Lennon once said that from the docks in Liverpool, you could sense the great continent on the other side of the Atlantic: "When you stood on the edge of the water you knew the next place was America."[1] The next place wasn't America, it was Ireland or Wales (depending on what direction you were facing), but of all British cities, Liverpool had the strongest connection with the United States. It felt its presence more than other parts of the United Kingdom, and its identification with the New World helped to define its place in English life.

A statue of Christopher Columbus in Sefton Park bears this inscription: "The discovery of America was the making of Liverpool." The city's special relationship with Americans, and their music, gave it a musical culture that played an influential role in the emergence of the Beatles. Many Liverpudlians will tell you that the Beatles could only have come from Liverpool with its unique mix of different cultures. Liverpudlians call themselves "scousers," after a stew that includes many different ingredients. As a great port city, Liverpool evolved out of the exchange of goods and ideas driven by the mighty engine of transatlantic commerce, creating a unique cultural heritage that helped shape its music. The trade links that joined the two continents dealt not only in hardware and recordings, but also in fabricating celebrity and exploiting the dreams and desires that sprang out of an admiration of all things American.

Back in its heyday in the nineteenth century, Liverpool was one of the busiest seaports in the world. It called itself "the Gateway to the Empire" at a time when the British Empire stretched across the globe. Its trade, shipbuilding, and other manufacturing industries made it the wealthiest city in England in 1800 and one of the most prosperous communities in the whole of Europe. Liverpool claimed to be a world city, rather than a provincial center in Lancashire in Northwest England. It was a town that looked beyond the River Mersey to the great outside world. At one point in its history it could claim more than a tenth of all the tonnage sailing the seas and a third of the world's trade. Its wealth was built on its trade with North America, especially the southern colonies, which began when boats from Liverpool brought back tobacco from the newly established Virginia colony.

In 1709 the first ship from Liverpool took fifteen slaves to the colonies, the beginning of what the English merchants delicately called the "African trade," which would last a century, enriching the city and bonding it to the plantations and prosperity of the American South. By the end of the eighteenth century one-third of the port's business was with North America, and this relationship was so important that Liverpool merchants protested loudly and often to the Crown about its harsh treatment of the American colonies in the 1770s. Liverpool's merchants took the cloth of Lancashire's textile mills, often the distinctive blue and green striped material from Manchester, to West Africa and the Slave Coast, where they exchanged it for valuable trade goods: "slaves, gold dust, and elephants' teeth."[2] The slaves were then taken to plantations in Virginia, the Carolinas and West Indies. The ships returned home loaded with rum, sugar, limes, dyestuffs, and other exotic products. Liverpool shipped in so much rum that it became notorious for the number of places selling liquor in town. London might have had its gin palaces, but every seventh house in Liverpool reputedly sold alcohol. It was, and still is, a hard drinking town whose people enjoyed a good time after a long day's work.

In the eighteenth century Liverpool competed with London and Bristol (a port on the west coast, close to Wales) for the domination of the slave trade. Liverpool won. By 1800 it accounted for 85 percent of the African trade, and the profits amounted to 1.5 million pounds each year. The last slave boat left Liverpool in 1807, with four hundred people in its hold. Although the slave trade was officially ended, Liverpool retained its dependence on "King Cotton," importing millions of pounds and distributing it to mills all over the industrial North of England, where production continued around the clock. Liverpool's wealth was built on cotton and the slave system that supported it, a global system of trade and exploitation that the Succession Convention of the Confederacy called "the greatest material interest of the world" as it announced its separation from the United States in 1860.

During the American Civil War, sympathy for the South ran high. The Birkenhead (Liverpool) shipbuilders Cammell Laird built the Confederate commerce raider *Alabama*, which was specially designed to prey on Union ships in the Atlantic. All in all Liverpool built thirty warships for the Confederacy, and the indignant Union envoy to the United Kingdom claimed that the city was "unofficially fighting on the side of the South."[3] (The official Confederate representative in Liverpool lived in "the Alabama house.") The merchants of Liverpool held charity events for the Confederate sick and wounded and lamented the effect of the Civil War on their trade. The city still bears some signs of its profitable commerce with the tropics. Many majestic homes of the wealthy merchant classes still stand in downtown Liverpool, and in the mass of ornamentation, you can see depictions of exotic plants and fruits (such as the pineapple) that were native to the South and the Caribbean. Liverpool's commercial elite felt no guilt about the trafficking of slaves; in fact European merchants rather envied Liverpool for the profits it made from it. Not until the mid-nineteenth century was some relief expressed that the city was putting up fine buildings that had not been founded on "the blood of negroes."[4]

Liverpool wore the evidence of its trading networks in different ways, and most noticeable were the African, Asian, and Caribbean sailors who took up residence in the city when not at sea. Liverpool's population was cosmopolitan. It had England's first black community in the 1730s, its first Chinatown, and by the 1800s it could claim to be the most diversified city in the UK. That is not to say that it was an oasis of tolerance and equality; anti-Irish sentiment often flared up in the Protestant communities, and there was certainly prejudice against Africans and Asians. Writing in 1907, Ramsay Muir, a professor of history at the University of Liverpool, commented on the polyglot nature of Liverpool and added that some of the races were "backward in many ways."[5] In the 1930s and 1940s, as Liverpool emerged as an entertainment center, an unofficial but rigid segregation was in force, and racial strife erupted in the 1960s and 1980s.

Immigrants poured into Liverpool throughout the nineteenth century. Many of them came from Wales, which sits just south of the city across the River Dee. The Welsh are a proud and independent people with their own melodic language and a long musical tradition. Among this group were the ancestors of Allan Williams, the Beatles' first manager; Alun Owen, who wrote the screenplay for their first film; and Annie Jane Millward, maternal grandmother of John Lennon. The stream of immigration became a flood after the great potato famine devastated the population of Ireland. Nearly 300,000 Irish peasants arrived in Liverpool during the famine, and about 130,000 of them kept on going to the United States. The remainder helped double the population of the city. By 1851 the Irish made up more than 20 percent of the city's population. Included in the thousands of newcomers were John O'Leannain and James McCartney—the kith and kin of the Beatles and their extended families. They crowded into dense residential neighborhoods often named after the places they left in Ireland, such as "the Dingle," where Ringo's family lived. The Liverpool Irish, like the East

India men, the Welsh, the Chinese, and the Africans, gave the city its cosmopolitan culture and contributed to the unique Liverpool scouser dialect. In the late nineteenth century, many more immigrants arrived in Liverpool from Eastern Europe. Among them were Brian Epstein's grandparents, who joined the growing Jewish community in Liverpool and opened a furniture shop in Walton, a suburb north of the city. There were districts where Irish and Jewish immigrants lived together. The entertainer Frankie Vaughan came from Devon Street, "a real Cohen and Kelly district, a place where the Jews and the Irish have rubbed shoulders for years."[6]

In the eighteenth century the networks of commerce and communication facilitated a transatlantic flow of people, products, and ideas. The invention of the telegraph and the laying of the transatlantic cable in the nineteenth brought the two sides of the Atlantic much closer together. The steam-powered railroad and steamship accelerated transatlantic trade and migration, cutting down a fifteen-week journey to two or three. Liverpool was at the center of this transportation revolution. It was connected to its neighbor Manchester by the world's first commercial railway—an installation that was frequently visited by American businessmen—and its shipyards built the steamships that continually cut the time to cross the Atlantic, including the sleek *Mauretania*, which did the crossing in only five days. An examination of the family trees of the Beatles shows how important seafaring was to the economy of the city and the lives of its inhabitants. John Lennon's grandfather was employed by a salvage company and was often at sea, while his father, Freddie, worked the passenger liners that crisscrossed the Atlantic and was rarely at home. George Harrison's father, Harry, worked as a steward on a transatlantic steamer. Liverpool was the hub of the transatlantic mail and passenger service and the home port of the Cunard and White Star lines.

These powerful links ensured a germinating flow of new ideas and skilled people across the Atlantic. The technology of the first Industrial Revolution traveled westward to America in the hands of mechanics and the papers of industrial spies. Dramatic improvements in the techniques of mass production moved back eastward in finished goods and exhibits for world expositions. By the last quarter of the nineteenth century, the speed of communication was increasing as fast as the flow of new products streaming from corporate and independent research laboratories. These "invention factories" produced so many dramatic new technologies that scholars have recognized a second Industrial Revolution, one of electrical power and motion pictures instead of steam engines and cotton cloth. The innovation moved from heavy industry across the spectrum of the economy, and brought significant changes not just to transport and manufacturing but also to consumer products and services that affected everyday life. Entertainment was mechanized like everything else—the old entertainments were modernized and new ones devised around amazing inventions.

THE MECHANIZATION OF ENTERTAINMENT

Our Victorian forefathers enjoyed a song or two while they drank. Music hall entertainment started in the tap rooms of pubs, when singalongs by the customers and "turns" by semi-professionals provided music to go along with the dancing and drinking. In Liverpool these places were called "free and easies" and earned a reputation for rowdiness. The Liverpool *Porcupine* lamented "the evil influence of dancing saloons."[7] Once established in small, poorly lit venues with little heating and no air conditioning, these entertainments were transformed by the technological advances of the late nineteenth century. The first theaters and music halls were built of wood and illuminated by candles and oil lamps placed in the wings and snuffed out between performances. Better construction methods and more durable (and fire retardant) materials allowed the music hall to grow larger. It was the nature of the business for new management to take over a venue, redecorate it, and bring in "higher class" acts. They replaced wood with brick, and candles with gas lamps. The Theatre Royal in Liverpool advertised more chandeliers and better lighting after it had been improved in 1831. By mid-century Liverpool had several large and ornate music halls, such as the Theatre Royal, the Adelphi, the Royal Hippodrome, and the Olympia.[8]

There was a music hall for every taste and budget. Liverpool offered the whole spectrum of pleasurable diversions, from the opium dens of Chinatown to the great palaces of variety entertainment in the city center. But Liverpool's reputation reflected the infamy of its cheap and shabby dance halls for the working classes. A Home Office report stated that no other city had venues that had such "a demoralizing influence" on the young. Yet even the big theaters had audiences that were often uncontrollable. A commentator at the Theatre Royal noted that a "more turbulent, indecent and tasteless audience" had seldom been seen as drunken sailors "and their doxies" filled the gallery. It was common for theater owners to erect a barrier between the orchestra pit and the front row to stop the mob from invading the stage.[9]

The owners developed a business strategy of raising standards, prices, and the class of people who were attracted to the show. This affected both the character of the entertainment and the place where it was presented. The more luxurious the venue, the higher the class of patrons, who could be persuaded to pay more.

The motion picture industry learned this lesson in the 1920s and 1930s as they too built "palaces" for their more affluent customers and moved well beyond charging five cents for a ticket. Modern plumbing and lighting was a major attraction. Theaters were among the first to acquire electric lighting at the turn of the century. The Queen's Theatre in Liverpool advertised its electric lighting in 1904, and several more large establishments followed: the Empire, Hippodrome, and Pavilion. The Liverpool Empire was one of the largest venues in the country when it was completed in 1904. It had seating for 3,750 and was part of the prestigious Moss circuit of theaters.

The traveling shows that brought music and dance to isolated communities from Deadwood, Wyoming, to Aberdeen, Scotland, had started in the horse and buggy era. Railroad cars, steamships, and later motor cars transformed them, enabling larger business organizations to trade across borders and rationalize their product. The railroad and telegraph took vaudeville nationwide in the United States by putting it on the rails. Minstrel shows, circuses, and Wild West extravaganzas incorporated hundreds of performers, assorted animals, and tents big enough to comfortably hold a thousand people. Managing these shows' transportation required advanced organization, communication, and technology. Behind the circuits of music halls were a few centralized booking agencies that monopolized the hiring and placing of performers, shifting them around the country as needed. Their office workers used the newly invented telephones, typewriters, calculators, and dictating machines to manage this sprawling operation.

The growing size of the entertainment and the ability to move it around quickly soon justified larger venues. The process of raising the capital to build much bigger and permanent structures also encouraged rationalization. Businessmen incorporated many small family-owned venues into national chains of theaters under one management—the Alhambras, Empires, and Palaces that stood in every large English city. Liverpool had over twenty music halls, ranging from small family-owned operations to magnificent examples of Edwardian splendor, such as the Hippodrome and Olympia, which showcased star entertainers and often brought in the well-known American acts, such as the escape artist Harry Houdini and the comedian W. C. Fields, to add to their prestige.

The transportation revolution made popular entertainment a transatlantic business that exploited common languages, communication systems, and railroad gauges. The popular Wild West shows are an example of what could be accomplished with improved technology and management. They began on the American frontier after the Civil War as a diversion for the tourists coming from the East. Railroads brought more tourists and gave entrepreneurs like William

Cody the idea that the same transportation system could take the shows to the East. Buffalo Bill's Wild West became the leader in these shows and created a mythology around Cody and the development of the frontier. The ambitious Cody and his backers saw opportunities in Europe, and in 1887 the show sailed for London with 209 performers (including a contingent of Lakota Sioux), nearly two hundred horses, and assorted buffalo, elk, and deer. This was entertainment on a truly industrial scale. Cody returned to Europe several times, traveling across the continent as far as Ukraine. He claimed to have logged 63,000 miles in all.[10]

Although many are nostalgic for the music hall of these good old days, it was in fact a vast, highly organized global network of entertainment—the first great shock of technology battering against existing folk traditions of music and dance. It was the high-tech entertainment of its times. The faint and often fragile cultural bonds of the eighteenth century—the links that lived on in timeless melodies or half-remembered stories carried by immigrants—were replaced with a constant intercourse of professional musicians and actors. The music hall was an international business, bringing the two English-speaking sides of the Atlantic together in common entertainment. It brought some of the great names of the American stage, including Julius Brutus Booth, and popular American melodramas, such as *The Virginian Mummy*, to Liverpool's music halls. Liverpool's status as the chief Atlantic port meant that English stars often gave a last performance in the city center before they embarked from Pier Head. Numerous American dramatic actors, comedians, and singers played Liverpool in the nineteenth century and treated Merseyside audiences to the latest in American theater. In 1867 the Bijou presented the American actor E. A. Sothern in a production of *Our American Cousin*, a play about the misadventures of an American in England. During a performance of this play in Ford's Theatre in Washington, D.C., two years earlier, John Wilkes Booth, the son of Julius Brutus Booth, assassinated President Lincoln.[11]

The minstrel show was the most influential form of American entertainment in nineteenth-century Liverpool. It claimed to be the song, dance, and humor of the antebellum plantation South but was in fact a crude, racist distortion that primarily benefited white entertainers. In 1840 the American comedian and singer T. D. Rice appeared in a play at the Royal Court Theatre. His role as Jumbo Jim did not impress the audience, but his "Jump Jim Crow" routine delighted them. Rice was a major figure in the rise of minstrelsy in the United States—the segregation laws in the southern states were named for this routine—and played an important part in its diffusion to Europe. In the 1850s and 1860s more and more minstrel shows came to Liverpool, including the famous Christy and Virginia

Minstrels. In 1865 Sam Hague formed his minstrel company in Macon, Georgia. A year later they were appearing at the Theatre Royal, Liverpool, as the American Slave Serenaders—the "only combination of genuine darkies in the world."[12] Minstrelsy was also the mainstay of St James Hall, which introduced it in 1870 and kept it on the bill for the next twenty-five years. Minstrelsy did not die in the twentieth century; rather it survived and prospered on radio and television. At the time the Beatles were playing their first gigs, you could watch minstrel-like entertainment on primetime television or see it in person on the concert stage in the United Kingdom.

Improved transportation and modern business organizations increased the exchange of entertainment between Liverpool and America in the twentieth century. In 1908 one of Fred Karno's troupes of actors and comedians put on a pantomime show in Liverpool's Olympic music hall.[13] Among the performers was eighteen-year-old Charles Chaplin. (Also on stage with him was an equally young beginner called Stanley Jefferson.) Chaplin had already played Liverpool many times before, appearing first in 1899 with the clog dancing act the Eight Lancashire Lads at the Empire and later at the Hippodrome. Chaplin had toured England playing small variety theaters, but the Karno operation was a cut above groups like the Eight Lancashire Lads. In 1909 Chaplin went to Paris in a Karno troupe, and the next year he went to the United States, where he toured for two years, traveling by rail to all parts of the country. This transatlantic trip was nothing unusual; by the time Charlie went to America, Karno had been sending troupes there for six years. Chaplin's father was also a music hall entertainer, and he too had toured the United States.

EMIGRANTS AND INVENTORS

Entertainers were not the only people exploiting the transportation revolution. Theater managers and song writers like Oscar Hammerstein were also crossing the Atlantic, along with a stream of skilled craftsmen, mechanics, and engineers. Thomas Edison's research laboratories were full of European immigrants, and they made significant contributions to important inventions like the phonograph, which emerged from his Menlo Park laboratory in 1877. Soon there were several competing machines on the market, including the graphophone, developed by the Scot Alexander Graham Bell, and the gramophone, invented by Emile Berliner.

Berliner had left Germany in 1870 and established himself in Washington, where he devised components for telephones and telegraphs and developed a microphone. Where Edison had used a cylinder to record sound, Berliner used a flat disc. This was easier to work with because master recordings could easily be duplicated by stamping out copies from a mold made from the master. At first these discs came with music on one side, the A side, but over time Berliner managed to record on both sides (A and B). Berliner and Eldridge Johnson, a craftsman and mechanic from Camden, New Jersey, formed the Victor company—the organization that would make "gramophone" the name for all talking machines, to Edison's consternation. Important inventions coming from recent immigrants to the United States should not come as any surprise, for the millions of Europeans traveling to more welcoming western shores brought with them skills and knowledge that helped make the United States the leader in industrial output and technology at the turn of the century.

Thomas Edison was so impressed at the enthusiastic reception of his phonograph that he set out to develop a machine that would do for the eye what the talking machine had done for the ear. He started a development program in his West Orange laboratory that produced the first motion picture camera and built the world's first film studio. From the 1890s onward the short but quite amazing films from West Orange captured the attention of the public. Edison had envisaged the motion picture as entertainment in the home—just like his phonograph—but the introduction of the film projector and the rapid construction of small shop-front theaters by alert entrepreneurs (many of whom were recent immigrants from Central and Eastern Europe) took the show into the public arena and made it a formidable competitor to vaudeville. Many music halls introduced projected film shows on the bill in an attempt to co-opt the new technology, but longer, narrative films and the creation of the film star ensured that motion pictures would stand alone in their own specially designed theaters. Charlie Chaplin jumped from vaudeville to the new motion picture industry in 1913, after his second Karno tour of America. He went from earning 4 pounds a week to 150. He was followed into the movie business by his colleague Stanley Jefferson, who later changed his name to Stan Laurel.

The phonograph and film businesses that Thomas Edison created in his laboratories were international in scope. Silent movies were entertainment products that moved effortlessly around the world and were intelligible to anyone who could watch a flickering film strip and use their imagination. European filmmakers exported their output to American theaters, and American silent films found a large audience in Europe. When Charlie Chaplin returned home in 1921, he was greeted by a public who knew his Keystone and Mutual comedies as well as any American.

Edison had already gained considerable experience in mobilizing capital for the international diffusion of his electric lighting system, and it was this global movement of finance that set the stage for the subsequent diffusion of other American technologies, like the phonograph and motion picture. He quickly set up phonograph companies and manufacturing facilities in Europe, beginning with the establishment of the Edison Phonograph Company of Great Britain (later the Edison-Bell company). His two main competitors followed him across the Atlantic. Emile Berliner created manufacturing plants in his homeland of Germany to build his gramophone, and the Columbia Graphophone Company (which had been established around the patents of Bell and his associate Charles Tainter) established branches in London and Paris.

The phonograph companies distributed products and ideas across borders, borrowed money in one city and set up factories in another, and sent their technicians and managers across the oceans. Much the same process helped the motion picture spread from the bases of its invention in New Jersey and Lyons, France, to the rest of the world. The movement of skilled personnel was critical in the diffusion of the phonograph, telegraph, incandescent light, and motion picture businesses. Fred Gaisberg worked with Berliner in his Washington laboratory, and the gramophone company sent him to London to supervise the building of their first recording studio in Europe. He became the recording engineer and chief talent scout for the English gramophone operation, and in this capacity he worked closely with gramophone subsidiaries all over Europe, from Spain to Russia. His recordings of the operatic tenor Enrico Caruso were the first best-selling records. After he made the master recordings in a hotel in Milan, Italy, he sent them to the record plant of the Victor company in New Jersey, where they were used to duplicate thousands of copies that were then sold worldwide through the Victor company's international subsidiaries. Gaisberg played a critical role in the spread of recorded sound. Among other achievements he introduced African American music to India. This was a time when determined and skilled individuals could build an international organization. The American businessman Louis Sterling went to London in 1909 to revitalize the graphophone business in Europe. He brought in new capital, rationalized manufacturing, and built up the roster of recording artists. He went on to play an important role in the diffusion of popular culture across the Atlantic.

American companies had begun to move into European markets at the end of the nineteenth century, especially with high-tech products like electrical generators and motion picture projectors. The first tour of Buffalo Bill's Wild West coincided with an American exhibition in London, which showcased American products and art. Americans had once marveled at the technology and culture of Europe, but now the tables were turned. The Big Three talking machine companies in the United States (Edison, Columbia, and Victor) did not stop after they had colonized Europe but soon expanded into Asia and South America. By World War I, there was hardly a place left on earth that had not been touched by the wonder of the talking machine.

The first phonograph boom produced a wave of startup companies to service the rapidly growing listening audience. Some were independent companies, others were subsidiaries of the Big Three, and some were foreign operations looking to establish a foothold in new markets. Among the latter was a group of companies owned by the Swedish inventor Carl Lindstrom, including Odeon (based in Germany), Parlophone (based in London), and the OKeh record label that had been founded in New York by the German businessman Otto Heineman. Entrepreneurs like Lindstrom and Heineman built business organizations with European finance and technology, but they recognized that American performers were the leaders of popular music and that an ambitious record company had to have a presence in North America.

The introduction of the Western Electric system of electronic sound recording in the late 1920s provided not only a far superior record—the 78 rpm shellac disc—but also the synchronized sound of the talking picture. The talkies brought the English and American film industries much closer, for not only did they share the same language, but the enormous capital requirements of converting to synchronized sound strengthened the financial ties between them. The Americans owned the technology and had the skilled technicians to make it work. This increased the dependence between the licensees in the United Kingdom and the research laboratories of the United States, and put the mantle of technical and creative leadership on the shoulders of the latter. The men who ran the entertainment business in Europe had to keep up with the innovations on the other side of Atlantic, which covered everything from new film genres to the strategy of selling food and drink in theaters, which made more money than selling tickets. In the first Industrial Revolution, Americans came to Manchester or Liverpool to learn about the machines that were changing their businesses. In the second, English entrepreneurs traveled to New York and New Jersey to review the new technologies they were licensing.

The high cost of the new electrical recording worked in favor of the larger companies and accelerated the rationalization of the English and American record and film industries, because many of the smaller, less capitalized companies could not afford the transition to electronic sound. The rough times of the

1930s dramatically reduced the number of small record labels and filmmakers. Only the strong survived and in most cases grew bigger as they bought up their competitors and integrated forward and backward. Aggressive entrepreneurs also consolidated the film exhibition business by absorbing the smaller locally owned theaters into great national chains, which were often owned by the film producers. This process began in the United States in the 1920s and in the UK in the 1930s, when the Rank and Odeon organizations in England built up their national networks, followed by the Gaumont and Paramount chains in the early 1940s.

The Great Depression helped form large international business organizations that dealt in entertainment. An executive of the British Decca record company, Ted Lewis, saw the Depression as the opportunity to acquire some American companies cheaply and expand his operations into the United States. After failing to get hold of the ailing Columbia company, he went on a spending spree, acquiring recording studios, pressing plants, and musical talent to form an American Decca organization that captured the largest part of the American pop music market by the end of the decade.

Louis Sterling of the UK Columbia Graphophone company was also busy taking over failing record labels. He did such a good job of building up his UK Columbia company that he could buy out its American parent in 1925; then Sterling acquired many European record companies as they struggled to make the transition to electrical recording. He merged the Parlophone, OKeh, and French Pathe record companies to make Columbia a force in the record industry. At the beginning of the 1930s the UK Columbia was running neck and neck with the European branch of the Victor company, His Master's Voice (HMV), to dominate the European market for popular music with their electrically recorded discs. When the Depression slashed record sales, Sterling grabbed the opportunity to acquire HMV and merge it with Columbia to form Electric and Musical Industries (EMI) in 1931. Sterling later sold American Columbia because he was worried about anti-trust issues in the United States, and it was eventually taken over by the American Record Company. ARC had been founded by Herbert Yates, who ran a film processing laboratory, to exploit the bargains in the American record industry. By 1932 ARC was the largest record company in the United States, but EMI was the largest company in the world. Based in London, where it had a large R&D department and a brand new complex of recording studios at Abbey Road, it did business in nineteen countries.

By the eve of the World War II, transatlantic popular entertainment was dominated by a few fully integrated, multinational companies that dealt in films, records, and sheet music as well as having interests in radio, booking agencies, and the press. Their management often came from film production and exhibition, but the introduction of the Western Electric system of electronic amplification concentrated economic power. This foundation in sound recording technologies gave these large organizations the means to dominate the record and film businesses. They were Empires of Sound. They operated within an international framework of finance, entertainment, and technology. Instead of the traveling shows and performers who brought music and dance in person, these integrated multinational companies distributed entertainment globally.

EXPORTING POPULAR CULTURE

American ragtime, minstrelsy, and syncopated dance music started in music halls and tent shows, but now they traveled the world in mass-produced records and films. Minstrelsy in music, dance, and humor quickly became the first mass entertainment of the twentieth century. Pretty soon there were English minstrel players, who did their best to imitate the dance steps, banjo playing, and evergreen songs of the minstrel show. John Lennon's grandfather Jack emigrated to the United States and joined Andrew Robertson's Colored Operatic Kentucky Minstrels in the 1890s. He became part of the transatlantic exchange of entertainment and appeared in Ireland in 1897. Returning to Liverpool in the twentieth century, he passed his show business expertise to his son Alf "Freddie" Lennon, who delighted in entertaining the family with impersonations of Charlie Chaplin. His wife, Julia Stanley, was also a natural entertainer, and her enthusiasm for the banjo and accordion was passed on to her son. Significantly, John Lennon's musical education began with an instrument strongly associated with the minstrel show.

Records brought American music to England and made American entertainers household names there. Freddie Lennon could do pitch-perfect imitations of Al Jolson and Louis Armstrong, two stars who grew up in the black-face minstrel tradition. The recordings of artists like Jimmie Rodgers brought the authentic sounds of the Deep South to every corner of the globe. George Harrison recalled that the first guitar music he heard came from a Jimmie Rodgers record. Gene Autry's "South of the Border" had the same transformative effect on eight-year-old Richard Starkey (Ringo Starr), sending shivers down his spine.[14] Imported American jazz records had a significant influence on popular culture in England in the 1920s and 1930s, bringing with them new dances, new clothes, and the hedonistic outlook that characterized the Jazz Age. Big band jazz dominated European entertainment from the Great Depression to the end of World War II. Paul McCartney's father, Jim, played the piano, especially the ragtime rhythms, the rolling, syncopated lines that filled the smoke-laden air in pubs and dance halls. He formed his own band, the Masked Melody Makers, with his brother Jack on trombone, and then Jim Mac's Jazz Band to play the dance halls of Liverpool.

Jim Mac's Jazz Band was part of an enthusiastic British embrace of swing music and dancing. Big band recordings, massive hits like Glenn Miller's "In the Mood," were followed by the movies and newsreels that showcased the fashions and new energetic dances like the jitterbug. The great American big bands did not tour Europe as much as the variety troupes they had displaced; moving so many musicians and their equipment was expensive, and the British Musicians' Union took a strong stance against American performers playing in England, insisting on a quid pro quo policy that severely limited opportunities to bring big bands to Europe. But this gave every incentive for British musicians to take up the slack, and Jim Mac's Band was one of the thousands formed to provide live music for the dances that went on every weekend. The English jazz bands did not attempt to alter the music or advance the artistic development of jazz; instead they slavishly copied the records and did the best they could to sound American. White musicians had been stealing musical ideas from African Americans since the very first jazz record was released in 1917. English musicians' attempts to mimic black vernacular usually failed miserably, but that did not stop them from trying and then handing down this particular musical tradition from generation to generation.

The war years brought England and the United States even closer and strengthened the financial and technological bonds between the two countries. Liverpool's merchant seaman maintained the lifeline with the arsenal of democracy and helped accelerate the Americanization of English popular culture. The most noticeable agents of this diffusion were the millions of American servicemen who arrived in the UK in early 1944 in the buildup to the great D-day invasion. They entered the country through the port of Liverpool, which had again become the fulcrum of the Atlantic trade. The docks were full of Liberty transport ships and the troop ships queuing up, twelve at a time, to disgorge GIs to Pier Head and to the trains that left every hour. Close to the city, the U.S. Army Air Corps (later the USAF) established the massive Burtonwood air base to service the armadas of American aircraft that flew over the continent every day.

The troops brought with them some of the luxury goods, like chocolate and nylons, that had disappeared from English stores years ago. They also brought chewing gum and Camel cigarettes, and plenty of money to spend. Although the locals were glad to be allied with such a rich and powerful nation, the "Yanks" were not always welcomed with open arms. They were "over-paid, over-sexed and over here," in the phrase made popular at the time, and there were plenty of hostile incidents, especially over interracial liaisons, and a growing sense of envy among deprived British males. Liverpool, like other cities involved in the war effort, was a hive of manufacturing activity that led to a dramatic increase in incomes and nightlife after the long hours of war work were finished. Overtime put money into the pockets of thousands of young women, and there was a never-ending stream of servicemen looking for distraction from the serious business at hand. The pubs and dancehalls were full of people desperate to have a good time while there was still time to have it. Attractive young women like Julia Stanley could go out every night and dance and drink until dawn. They smoked American cigarettes, did American dances to American music, and communicated with newly acquired American slang. When they got tired of dancing, they could go to the local cinema and enjoy a Hollywood movie.

The war years established the dominance of American big band jazz in Europe because all the combatants danced to swing, even the Germans. But the introduction of a million Americans into England also brought other sorts of music, and different regional accents, to the attention of the English. The Armed Forces radio Network (AFN), established in 1942, broadcast American music to all corners of the globe and gave foreigners the chance to tune in and hear something completely new. This extensive radio service was part of the U.S. government's strategy to bring American culture to GIs serving overseas—the plan that delivered Coca Cola and American newspapers to troops stationed on every front of the world war. By 1945 AFN had eight hundred stations worldwide. It brought the music of African Americans to far-flung listeners, and although the American armed services were officially segregated, many white soldiers got their first taste of rhythm and blues from it. The great swing bands like the Count Basie Orchestra entertained black servicemen all over Europe, but smaller combos swung a little harder and louder—the jump jive of groups like Louis Jordan and His Tympany Five. This was the origin of the rhythm and blues designation, which replaced the "race music" label that was invented by the record companies to market their products to African Americans. Even blues and gospel were broadcast to places where the segregated context of their production did not hamper their reception. More than one soldier or sailor went home after the war with a new appreciation of black music and culture, and more than one city in Europe and Asia finished the war with the sound of the American South ringing in its ears.

The rich musical tradition of Liverpool drew its energy from its long association with American popular culture in all its many forms, but it was seasoned with the obstinately scouser vernacular and the powerful heritage of the English music hall.[15] The Beatles absorbed it all. They went to America as legitimate ambassadors of Merseyside culture, representing a city that had a special affinity to African American music. Well before the Beatles performed at the Cavern Club in downtown Liverpool, blues artists like Sonny Terry and Brownie McGhee had already played there to an educated and appreciative audience. Giants of modern jazz Theolonius Monk and Art Blakey were also invited to Merseyside to entertain their many fans. The links between Liverpool and American culture had been forged well before the Quarry Men emerged, but it was the Beatles who consummated the relationship in the final act of this transatlantic love affair.

CHAPTER 4

THE PROMISED LAND

For the Beatles and the postwar British youth they represented, the United States was the promised land, the dream world of plenty that stood out in brilliant Technicolor against drab and dreary 1950s England. By the time the Beatles were teenagers, the country had still not repaired the damage caused by World War II, and it was struggling to reach prewar standards of living. Pop singer Marty Wilde remembered that black, brown, and gray were the colors he associated with the war and its aftermath.[1] Rationing was still in force, and the drastic shortages of food, energy, and opportunities had become a way of life to the weary population. For cities like Liverpool that had taken the brunt of the German bombing campaign, the signs of postwar stagnation were everywhere, in the burnt-out buildings, the lines of housewives queuing up for precious sugar or bananas, the Victorian outside toilets, and the whole street blocks of rubble that remained uncleared. The aftermath could also be experienced in the gloomy outlook of the British people, who had given their all in the war and found little in the peace to raise their spirits, as one crisis—heating, sterling (currency), and strikes—followed another.

The young often found escape in the imaginary worlds of sports, music, and film, with the latter the most powerful in illustrating the limitless opportunities of the United States. As Pete Frame confessed, it had "the best of everything, so we believed," the best cars, stereos, telephones, and recording studios. Amateur musicians like the Beatles admired gleaming American guitars that "looked sensational to us. We only had beat up, crummy guitars."[2] Architects and urban planners were impressed by the futurism of American buildings and suburban developments. English kids saw the huge cars and amazing stereos in films and on television. Teachers and poets praised the boldness of a new generation of American writers, and schoolboys reveled in the rebellion of their words. Liking things American was more than an attitude: it was a movement. Frame concluded, "Worship of America and all things American was an established religion amongst the young and progressive."[3] America had grand canyons and imposing skyscrapers. It had the famed nightlife of New York and San Francisco, and at least one member of the Beatles' party was fascinated by stories about the subterranean gay life of American cities. "America was the place we all wanted to be," author Ray Gosling commented, speaking for the postwar generation of youth who looked at the United States with admiration and awe.[4]

But most of all, America had the best music, the most famous venues, and the biggest stars. It represented the highest peaks of professional attainment, the "eldorado of the entertainment world," in the words of George Martin. But it also threatened the lowest lows of potential failure. Perhaps Brian Epstein captured this mindset best when he said "Always America seemed too big, too vast, too remote and too American."[5] This was playing on the mind of Paul McCartney during the flight to Kennedy Airport in February 1964, when he thought out loud: "They've got everything over there, what do they need us for?"

England in the early 1960s was particularly in thrall with the United States, whose industrial output, technological innovation, military power, and political vigor were the envy not only of the British but of most of the noncommunist world. In the aftermath of World War II, the rest of the twentieth century was clearly now in American hands. In 1960 John F. Kennedy became president, and his youth, ambition, and optimism personified the promise of America to many foreigners. He was the first president to appreciate the power of television, and the story of his presidency, from primaries to Dallas, was enacted on television screens. He defined his presidency in terms of a new generation of young Americans and their unbounded belief that they could change their world. England's left-leaning Labor party, under the leadership of Harold Wilson, looked to Kennedy's success and tried to appropriate some of the new marketing and organizational ideas coming across the Atlantic. Labor's vision of "New Britain"—young, bold, creative, and proud of itself—took a page right out of Kennedy's playbook, as did Wilson's promise to produce growth and prosperity with the "white heat" of technological development. Wilson based his successful New Britain election campaign of 1964 on Kennedy's 1960 New Frontier campaign: aimed at youth, full of the promise of change, and powered by television.

In the 1950s the impoverished British had to retreat from their position of world power to hide behind the U.S. nuclear shield. To keep face, Britain was allowed its own nuclear weapons—but they were all obtained from American stockpiles and incorporated into imported American weapons systems. In nuclear technology, in the science of the space race, and in the everyday consumer products that offered a more comfortable life, the Americans were in the lead.

This was true of the social sciences as well as the applied hard science of the space missions. Plausible new gurus like Vance Packard and Timothy Leary created captivating new ideas in psychology and sociology that attracted England's well-educated teenagers. And as much as Yankee ingenuity impressed the rest of the world, it was the Yankee ability to turn technology into money that European businessmen envied the most. New ideas were the work of not only rocket scientists but also psychologists, anthropologists, and sociologists. They applied their ideas to commerce, especially advertising, marketing, and public relations, and these businesses expanded rapidly in the United States in the 1960s, when advertising billings increased more than 75 percent in the first half of the decade. Advertisers had bright new designs and ideas, often crass but always engaging, that blew into every dark and cobwebbed corner of English industry. Advertising and public relations were examples of the new, modern businesses of the 1960s, and they did much to create the decade's glossy image.

American leadership was not only in organizational, industrial, and scientific knowledge. While Harold Wilson and his government were buying American Polaris submarines, English schoolboys and girls were avidly reading *Catch 22* and discovering Ralph Ginsberg and Jack Kerouac. Americans seemed so modern, and their literature, poetry, and art reflected this sense of newness and originality. The beat poets and brilliant new books aimed at youth, such as *The Catcher in the Rye*, mesmerized British students and intellectuals. All over the country, young men and women gathered in coffee bars, pubs, and classrooms to discuss new American literature and poetry, just as John Lennon, Stuart Sutcliffe, and Bill Harry were doing in Liverpool. Bill Harry recognized that the American music, films, books, poetry, and comics seemed far more glamorous than their British counterparts.[6] The British art world looked up to the modernism of America, and the practitioners of the newly named pop art saw the United States as the benevolent font of ideas. Artistically inclined young Englishmen thought that the music, art, and literature that really counted came from the United States: "It was as if the U.S. was our spiritual home," said Tony Bramwell.[7]

AMERICA ON FILM

British views of America were formed first by novels and comics and perhaps a visit to Buffalo Bill's Wild West for those lucky enough to get tickets for its European tours. Then motion pictures turned these words and memories into stirring, mythological images. Silent films could travel unhindered by the barrier of language and bring to life a powerful vision of America and its history. In the 1930s synchronized sound brought the language, the rough but entrancing vernacular of cowboys and gangsters, to English-speaking viewers. The talkies made Hollywood, and they taught American history and culture to a generation of young Europeans.

Hollywood had a strong foothold in the UK because of financial ties and a common language. Opulent cinemas in London's West End, the entertainment center of the capital, were copies of the great picture palaces in New York City. In the rest of the country, film theaters large and small showed American movies. As Hollywood reached its golden age in the 1930s, it set the standard for cinematic art and commerce. Subsequently American films filled large parts of the bill of the average British cinema, from the Saturday morning Westerns that continually recycled B movie footage shot in the 1930s, to the main attraction of the week, a Hollywood blockbuster like *Gone with the Wind* or *Casablanca*. Europeans watched American films of every denomination, from *My Man Godfrey* or *Mr. Deeds Comes to Town* to Bing Crosby in *Rhythm on the Range*, which all appeared on English screens in 1936. In that year alone English audiences could see slices of life from many parts of American society: the rich, the poor, and even African Americans, whose depiction in *The Green Pastures* was as frivolous and fanciful as any filmed stereotype. As soon as the Beatles were old enough to go to the pictures, they were bombarded with images like these.

Like the Beatles, director Terence Davies grew up in Liverpool after the war. He recorded his impressions of youth in his elegiac film memoir *Of Time and the City* (Hurricane Films, 2008). As he tells the story of his childhood, he reveals that American films had a profound influence on him from an early age, not just drawing him into a distinguished career in filmmaking, but also influencing and perhaps molding ideas of his own identity. Films obviously brought glamour and excitement to dreary, everyday life in Liverpool, but they also plumbed deeper and more personal currents of imagination. The world that Davies saw on screen was much different from the world outside the cinema; it was "glorious old Hollywood" instead of "small, comic England." He affirms that his generation saw America as magical land, a place where everything seemed perfect. Davies uses his own film, described as "a love song and a eulogy," to lament the passing of the Empires, Scalas, Ritzes, and Gaumonts that opened his window to another world across the Atlantic.

The Beatles felt the pull of America well before the advent of rock'n'roll. They had seen the same films as Terence Davies and perhaps dreamed some of the same dreams. The movies they saw contained powerful mythic images, and as television developed in the UK, especially independent television in the mid-

1950s, many more of them filled British popular culture. The power of these images to capture the imagination of English youth was revealed in the Davy Crockett fad in 1956, which swept up male adolescents in an orgy of anglicized recreations of the Western frontier, when a generation of boys put on the coon-skin caps and made every park and piece of wasteland in the kingdom a potential "wild frontier." The Davy Crockett character was the creation of the Walt Disney company, which was experienced in bringing together sound and image to create lovable characters, especially Mickey Mouse, and then merchandising them worldwide. Disney made more money selling Mickey Mouse toys than it did from the cartoons. Even with this experience behind them, Disney was astounded by the success of Davy Crockett. Three hour-long episodes on the weekly *Disneyland* television show took the American public by storm and made Crockett and the actor who played him national heroes. Forty million Americans watched the shows, and millions more bought tickets to the feature film that was quickly assembled from the episodes.

A nostalgic return to the Western frontier had obvious appeal to Americans in the 1950s, but more surprising was the show's effect on foreign audiences. The stories were infused with a mix of patriotism and self-reliance typical of the American media at this time. The boy-scout homilies, "Be Sure You Are Right—Then Go Ahead," placed Crockett squarely in the frontier tradition of benevolent "cowboy codes" (a far cry from the reality), yet they managed to capture the imagination of Europeans. Disney exported Davy Crockett to thirteen television markets overseas and did phenomenally well in all of them. Audiences all over the world embraced the theme song of the show because sixteen different versions were recorded. Millions of boys sang "Daveeeee, Davy Crockett, king of the wild frontier" as they played with their Crockett hats, belts, plastic knives, and "Ol' Betsy" rifles. Davy Crockett proved to be a more effective merchandiser than Mickey Mouse. About two hundred Crockett items generated $100 million for Disney, including over 10 million coonskin hats.[8] It seemed like every boy in the United Kingdom had a Davy Crockett hat, and all of them were singing "The Ballad of Davy Crockett." If the record industries on both sides of the Atlantic needed any confirmation that television exposure sold records, this was it. There were three cover versions of this song in the British charts in 1956.

These events showed how the international interconnections of film, television, and record companies could transport a cultural phenomenon quickly and efficiently across the Atlantic, turning it into a self-sustaining brand. When Wim Wenders, the German film director, talked about how American movies colonized the European subconscious, he was not thinking about Davy Crockett, but the craze proved his point perfectly; the "tales of the wild frontier" helped articu-late a mythical America among the European baby boom generation. Crockett's tall tales of bear hunting and Indian fighting mined the vast reservoir of images of the West that went back beyond Buffalo Bill's show. Even though Davy Crock-ett's adventures were far from realistic portrayals of the Western frontier (they were filmed in Appalachia), its young fans in Europe never questioned that this might not be the real thing.

Crockett joined a long line of irresistible American heroes who captured the imaginations of Europeans. Teenagers like the Beatles spent their impression-able youth watching Hollywood films and saw Americans as larger than life. Their celluloid heroes were much more real and animated than the stoic heroes of the war movies popular with British filmmakers in the 1950s, or the downtrod-den characters of the romantic comedies. English youth yearned to be Robert Mitchum or James Dean rather than English film stars like Kenneth More or George Formby. Like their friends and relatives, the Beatles loved Formby and delighted in his typical Lancashire humor. They laughed at all his jokes, but they did not put his photo on their walls and imagine that one day they might be like him.

American films had the most influence on the young. The circumstances of the postwar movie industry had pushed producers into catering to the youth market, because family viewers were lost to television. Film studios rushed to turn out teen product that was more appropriate to the suburban drive-in than the mighty Roxies and Paramounts of city centers, which were now only half full. They welcomed the rise of juvenile delinquency, outlandish fashions, and loud, offensive music that provided the subject matter for their films. At first the mu-sic wasn't that important; the draw was teenage delinquency and other unsocial behavior. Hollywood had figured this out after the runaway success of Marlon Brando in *The Wild One* (1954), the film that launched a thousand motorcycle gangs in England. Although all eyes were on the bikes and the leather jackets, the music slowly edged its way into films as rock'n'roll became more popular. It pro-vided the background for the drag-racing scene in the James Dean vehicle *Rebel Without a Cause* (1955) and an exuberant interlude in a sex-and-scandal story like *The Girl Can't Help It* (1956), when a short performance clip of Little Richard temporarily distracted attention from the curves of Jayne Mansfield.

Then MGM pictures made *Blackboard Jungle*, a film about juvenile delinquents in an inner-city school whose musical tastes clashed with those of their teachers: it positioned cultivated, cool jazz against loud and belligerent rock'n'roll. The Beatles and their peers had seen a few American teen movies by 1956, but none of them was ready for *Blackboard Jungle*. The opening credits were backed by a song from Bill Haley, a retooled country and western singer who had moved into faster and more danceable music called rockabilly. He had already had a hit in 1954, with a song copied from R&B singer Joe Turner called "Shake, Rattle, and Roll," but the other song he recorded in that same Decca session would transform his career and the popular music industry in Great Britain. "Rock Around the Clock" played for only a short time at the very beginning of *Blackboard Jungle* (and Bill Haley and the Comets were not credited in the movie), but its effect on the audience was electric. It sounded very loud on large cinema loudspeakers, the expensive American-designed systems that had brought dramas like *The Ten Commandments* to life, but this wasn't the sound of floods or temples crashing down—this was pure excitement, and it swung, really swung, with an urgent, pounding beat produced by a dynamic rhythm section of slap bass, guitar, and drums that complimented Danny Cedrone's biting electric guitar. A few startling minutes of this one song announced rock'n'roll to English youth and to the nervous older generation, who now saw their worst fears about teen music and juvenile delinquency realized.

Some cities in the United States banned the movie, perhaps frightened by its promise of "The Most Startling Picture in Years," but this was nothing compared to its reception in the United Kingdom. Maddened by rock music and emboldened by celluloid rage, gangs of English youth slashed the seats in cinemas and marched arm in arm down streets, shouting "We Want Rock!" This was the beginning of the end as far as the British establishment was concerned. Their fears about teenage delinquency and the pernicious influence of imported antisocial films, music, and comics had come true. But for many teenagers, *Blackboard Jungle* was a turning point, a coming-out ceremony into a new world of excitement and consumption.

The unexpected effect of *Blackboard Jungle* and its tumultuous music turned on the green light in Hollywood production offices. Several more rock movies starring Bill Haley followed: *Don't Knock the Rock* (with songs performed by Little Richard, Jimmy Ballard, and the Treniers) and *Rock Around the Clock* (with Freddie Bell, Tony Martinez, and the Platters.) Both were cheap, exploitative teen flicks hurried out of Columbia studios in 1956. The film's producers tried to reassure the older generation about the threat of rock'n'roll mayhem: "A fun loving group of kids moving to a new kind of beat they call rock'n'roll . . . Arnie and his gang hope to prove to all the adult cats in town that rock'n'roll might be really crazy, but it's not as dangerous as it looks."[9] Nevertheless, it was dangerous music, and each titillating burst of it reached deep into the hearts and minds of English listeners. Fifteen-year-old Gillian Shephard watched *Rock Around the Clock* in a quiet seaside town with a friend: "Suddenly she began to scream. I couldn't believe it. I was so embarrassed. She'd read about people screaming at the film and cinema seats being torn up."[10] The press jumped on the story, and widespread reporting of the effects of *Blackboard Jungle* brought in many impressionable teenagers as part of the experience—much the same way the press would propel Beatlemania a few years later.

Gillian Shephard was able to see American rock'n'roll movies in her hometown, as did the Beatles and all the other music fans who lived far away from London. Usually teenagers like these were excluded from the important developments in British popular culture, but the national circuits of cinemas brought it to them as quickly and as easily as they had brought Hollywood stars to their parents. Films about the effect of rock music on American youth rather than the music itself heralded the appearance of rock'n'roll in the UK. Films made Bill Haley the first rock star in England, and powerful images went on to make Elvis Presley the focal point of the new music.

The BBC did not play Elvis's records in the beginning, but newspaper pictures of him showed him in mid-gyration, guitar strapped around his body, legs spread provocatively apart, as he performed in front of a mass of aroused young women. Elvis Presley the performer had crossed over from country singer into mainstream entertainment before "Heartbreak Hotel" started slipping down the charts. No one can dispute that he made outstanding records, but what turned "Heartbreak Hotel" into a spectacular hit was its continual exposure on nationally broadcast variety shows in the United States. You could argue that Elvis Presley's material was not much different from the other white rockabilly singers, but where Elvis had the edge over Bill Haley and Gene Vincent was in his unmistakable good looks and his astounding stage show. Haley and the Comets put on a rollicking good performance, throwing themselves about the stage in routines they probably copied from the jump jive groups, who used dance moves to add spice to their act. Elvis, on the other hand, modeled his show on the R&B players he had seen in the dives of Memphis's red light district—singers like Wynonie Harris, who injected lascivious sexual suggestion into his performances. Elvis' pumped-up pompadour hairstyle, his colorful clothes, and his aggressive stance, with hips thrust forward, all came from African American acts. In racially sensitive America, Elvis's show was provocative and meant to be, but the condemnation that followed was grist for the publicity mill. His appearance on *The Ed*

Sullivan Show might have scandalized many American viewers, but the publicity made him a star and powered these explosive images across the Atlantic into the grateful grasp of the English press.

Films and photographs propelled Elvis and rock music into the consciousness of English teenagers. Wally Ridley had the job of building up HMV's pop music catalog, which in the 1950s comprised mainly licensed American recordings. Steve Shoals of RCA sent him "Heartbreak Hotel," and Ridley decided to issue it even though he could not make out the lyrics. The record was panned by the critics for being indecipherable and probably obscene, and it failed to get any radio play. Ridley was sure the record was dead when a few months later the *Daily Mirror* published a two-page spread on Elvis, and "the whole Presley thing suddenly broke over here."[11] His image was everywhere, and it was certainly a startling new look for English teenagers. Elvis had a brooding, dangerous air that definitely appealed to the boys, and his hairstyle and clothes, especially the gold lamé suits, made him look completely different from the homegrown singers he was knocking out of the charts. The Elvis brand was built on his looks. English kids could watch his moves on cinema newsreels, noting his tortured facial expressions, his sensuous moves, and the effect they had on the girls in the audience. They absorbed the images of screaming girls, hordes of excited fans, and the wild abandon of an audience caught up in excitement. Here are the roots of fan behavior that blossomed a few years later as Beatlemania.

After television appearances, Presley moved quickly into feature films, which appeared concurrently with the release of the songs he mimed on film. Elvis started on *Love Me Tender* in the summer of 1956, as "Heartbreak Hotel" moved up the English charts. The next year he made *Loving You*, which actually showed him performing as Elvis. *Jailhouse Rock* followed and established Elvis as a powerful presence in British popular entertainment. The film depicted the bad boy image that had become associated with rockabilly and some of the truly dangerous people who played the music. It was also a perfect articulation of the rock'n'roll mythology of rags to riches. Elvis played a working-class ex-convict who achieved almost instant fame and fortune in the entertainment business. It is easy to see what teenagers like John Lennon saw in this characterization of the music and its conflicting images of revolt and royalties. Elvis never played a concert in the UK, although so great was the yearning of his fans to see him in person that stories circulated of an impromptu performance during a refueling stop at a Scottish airport on the way to do his national service in Germany. But anyone could watch him gyrate through his hit songs on the big screen in the local Gaumont or Granada.

LOOKING AMERICAN

English audiences heard rock music as new and different, but what struck them most was the look; Elvis looked so amazingly exotic that one English commentator said it was as though he had just stepped off a spaceship.[12] We can see the influence of Elvis Presley on the Beatles in photographs of John Lennon at the Liverpool College of Art in 1957. While the rest of the group are in sensible sweaters and jackets, Lennon has the Elvis hairstyle that marks him as a rock'n'roller. In his bedroom he had cut the sides of his hair short and brushed it back while combing the long locks on top forward into a prominent "quiff," which required some sort of gel fixative, like Vaseline or Brylcreem, to hold it in place. This style was not Elvis's alone, for many other teen idols sported similar quiffs, such as Tony Curtis, who appeared in several American films (*The Sweet Smell of Success* and *The Defiant Ones*) shown in England at the same time Elvis's and Bill Haley's records were being played. At the rear of this hairstyle was another gel-coated appendage—the upwardly mobile DA (for ducks' arse), which did for the back what the quiff did for the front.

The size and architecture of the quiff signaled a level of Americanization and rebelliousness. It could get you into trouble at school, but it also gave you street credibility with the toughs of Liverpool. Each of the Beatles bought into this unmistakable fashion statement as they began to think of themselves as musicians. John Lennon's future wife described him as "the last stronghold of the Teddy Boys," the name given to kids who tried to look like American rockers.[13] George had a spectacular quiff that had to be kept up by liberal application of Vaseline, but this was beaten by Pete Best's towering hair, an important part of his good looks (often described as "brooding") that appealed to the girls on Merseyside. Stu Sutcliffe also had a quiff when he first joined the band, and his resemblance to James Dean (accentuated when he wore American Ray-Ban dark glasses) was one of the reasons he was in the band, because he could hardly play his bass guitar.

Music alone cannot explain the narcotic effect of American rock'n'roll on English youth; it was also the look, the attitude, and the language. These critical pieces of information came embedded in American films and photographs. Although every fan of rock'n'roll could impersonate an American voice, sounding American was not as important to teenagers in the UK as looking American: "That's what you aspired to, everyone wanted to look like a Yank."[14] During the Beatles' early days in Liverpool and Hamburg, they described themselves as "rockers" and tried as best they could to duplicate the look of early Elvis and the other stars of rockabilly—these were the kinds of Americans they yearned to be.

The look of the Beatles in the late 1950s reveals the gradual diffusion of American fashions that came over with the music. During the time they spent in Hamburg, they collected quite a few signifiers of the rockabilly style that was now associated with rock'n'roll: leather jackets, American jeans, cowboy boots, and the flat caps they copied from Gene Vincent and his backing band the Blue Caps. As Paul McCartney admitted, "We were four little Gene Vincents really."[15] They did not learn about Vincent from reading the musical press or examining his records; they hung out with him in Hamburg and learned firsthand how tough a rocker could be. The Beatles—like every other European teenager who wanted to be hip—looked to the rock'n'roll images coming from the United States as a fashion guide, from Ray-Ban sunglasses to cowboy boots. One of their friends in Hamburg, the photographer Jurgen Vollmer, described the look: "Everybody looked tough . . . Menacing looks. Black leatherjackets and hair styled into pompadours and ducktails were everywhere."[16]

While the British copied the music note for note, they added some of their own fashion heritage to the American look. They sought out throwbacks to Edwardian styles, such as the long jackets with velvet collars, and invented other elements, such as the "brothel creeper" thick-soled shoes that made up the exclusively British Teddy Boy (or Ted) look. This was begun by the young working classes of London in the early 1950s. They adopted styles from many quarters, including London's homosexual milieu and American Western films, and brought them together in an amalgam of rebellion and petty crime. The Teds were dandies, but they were also dangerous. They soon haunted cafes, clubs, pubs, and theaters across the country, and in the North they were a constant presence (and threat) in the venues played by the Beatles. Teds owed a lot to the biker culture beatified in *The Wild One*, and their paradigm of toughness was built around behavior they had watched in American films.

The Teds were devoted to rockabilly and immediately threw their support behind rock'n'roll, becoming a loud and highly visible part of its initial audience—they were largely responsible for the riots that followed *Blackboard Jungle*. Their allegiance was strictly limited to the early stars and their country and rockabilly roots, for the Teds' musical culture was firmly fixed in nostalgia for the 1950s and stayed that way for the next fifty years. In a development that can only be described as prescient, the Beatles soon left their Teddy Boy fashions and musical preferences behind as they began to absorb new ones, especially when they started to travel away from home. Unlike many of their musical peers in Liverpool, the Beatles were far more open to new influences and were willing to move with, or ahead of, the times. This they did throughout their musical career.

Although they thought of themselves as rockers and liked to indulge in fantasies of looking like tough guys, especially John, who Jurgen Vollmer described as quintessentially rocker (cool, aloof, but with aggressive restraint), they were certainly not into the violence or the gang culture that characterized the Teddy Boys. Vollmer caught the secret of the Beatles' appeal when he characterized them as rockers but with a foot in the other camp of the more sensitive and cultivated "exis," the artistic and enlightened young Europeans who were caught up in the romance of existentialism and other fashionable philosophies, and who listened to cool jazz and classical music. Vollmer described the Beatles as somewhere between these groups, "rockers by looks and exis at heart."[17] This ability to be all things to all people, to maintain the essential rock'n'roll paradox of tough exterior/tender interior, appealed mightily to teenage girls. The Beatles still kept enough rocker in them to impress Brian Epstein as "uncouth and ill clad" when he first saw them playing at the Cavern Club in Liverpool in 1961, but as he admitted later, something else drew him to them.

THE CUNARD YANKS

The American teen films distributed in England brought many more musicians to the attention of English teenagers. The plot of *The Girl Can't Help It* might have been as flimsy as Jayne Mansfield's clothes, but in addition to Little Richard, it presented performances from Eddie Cochran, Gene Vincent, Fats Domino, and the Platters. All these stars and all this music (in deluxe color too) made this particular piece of Hollywood fluff immensely popular with rock-crazed youth in England, including John Lennon and Paul McCartney. Once the two had been introduced at the church fete in Woolton in 1957, they circled around one another, getting a feel for each other's musical skills and tastes. Paul played John the song that introduced Eddie Cochran to both American and English audiences, for *The Girl Can't Help It* was the break Eddie needed to start his career. Paul knew the chords to Cochran's "Twenty Flight Rock" and all the words too. Lennon was impressed with his version of Cochran's song and invited McCartney to join the Quarry Men.

American film producers included popular African American performers in their teen films, which gave British youth a chance to see Little Richard or the Coasters and started many on a journey of discovery into African American music. They began by seeking out the recordings and then researching the origins of the music by the familiar process of show and tell, and the cross-questioning of record shop employees. The Beatles and their peers discovered the prehistory of rock'n'roll and the vital role played by the independent record companies in popularizing rhythm and blues. The shortage of shellac during World War II forced the majors into concentrating their resources on the top sellers in their pop and easy-listening catalogs, which left a window of opportunity for new companies to meet the needs of underserved audiences with country and R&B. A wave of new entrants transformed the American record industry immediately after the war, and independent record companies discovered some of the major talents of 1950s popular music, from Elvis Presley to Fats Domino. The Sun label's recording studio in Memphis, Tennessee, where Elvis was first recorded, now stands as a tourist attraction: the place where rock'n'roll started.

Although we generally look to the independent record companies in the Deep South, such as Sun and Dot, as the originators of rock'n'roll, the movement actually went on nationwide, and many significant records came from independent companies in postwar Los Angeles, such as Jukebox, Specialty, and Modern. Taken together, the output of these Los Angeles companies provided the Beatles with some of the most cherished records in their collections, and if you ever wondered where Paul McCartney got some of his distinctive vocal effects, especially the whoops and shouts that drove the fans wild, all you have to do is listen to some Little Richard records on the Specialty label.

The path that these precious recordings took to Liverpool has become a central part of the Beatles story. These inspirational R&B records were magically smuggled into Liverpool by merchant seamen, the Cunard Yanks, and they were the seeds that germinated into Mersey Beat. The Beatles' friend and assistant Neil Aspinall explained how merchant seamen brought over "a lot of American records that weren't being played in England, whoever found it first claimed the song as theirs." The liner notes on the back cover of *Meet the Beatles!* pointed out that Liverpool was the "hippest musical spot in the United Kingdom" because "its seamen bring the latest hit singles from America."[18]

The employees of the Cunard and White Star lines had a special place in Liverpool society. In the 1950s about twenty-five thousand of them worked the ships and lived lives that appeared far more exciting than humdrum jobs in factories or warehouses. They returned home from long periods at sea with pay packets in their pockets and presents from far afield. Their stories of New York and New Orleans, their flashy new clothes and Americanized language, earned them the nickname Cunard Yanks in Liverpool. They went to the United States for ten days and returned with "an American accent, loud ties and a baseball cap."[19] They have a special place in the history of Merseyside music, playing a significant part in what scousers think made Liverpool special. They have been credited as instigating the beat boom in the 1960s that fostered hundreds of new rock groups, including the Beatles. The Cunard Yanks were the key to unlocking the secrets of American popular culture to those who stayed at home, and as such, they were agents in the Americanization of Liverpool—the vital link to the promised land. In his biography of the Beatles, Philip Norman describes the Cunard Yanks bringing precious records to Liverpool made by unknown young African American musicians and concludes that while England listened to pale, homegrown imitations of American rock music in the 1950s, "Liverpool listened to rhythm and blues."[20]

Spencer Leigh of Radio Merseyside thinks this is "a good romantic story" that has endured for decades. It reflects a pride in Liverpool and its music, which often acts as a counterbalance to the unfavorable views of scouser culture shared by the rest of the country. The Cunard Yanks brought in the rare American records that gave Liverpool musicians the edge over players in other cities: "We used to get the soul records and rock'n'roll records long before anybody else."[21] Prem Willis-Pitts' account of the music scene in Liverpool also reflects this sentiment. He even claims that Merseyside groups were moving ahead of the Americans in covering Jerry Lee Lewis's rockabilly classic "A Whole Lot of Shakin' Going On" before it caught on stateside. Most contemporary accounts of the Cunard Yanks have them bringing in soul and R&B records, which strengthened the emotional ties between Liverpool musicians and African Americans.

Unlike their parents, who had drank and danced in segregated clubs in the 1930s and condemned socializing with black GIs in the 1940s, the generation of Willis-Pitts and McCartney was much more favorably disposed toward other races. Liverpool teenagers often revered black music and enjoyed good relations with the Afro-Caribbean immigrants who arrived in the early 1950s. These teens saw the emergence of the beat boom and the beatification of the Mersey Sound in terms of solidarity with African American musicians. The Cunard Yanks carried the records that held "the spark of life that sprang up in the black ghettos of America and leapt across the Atlantic to ignite the Liverpool beat."[22] So when musicians and journalists considered the Mersey Beat boom of the 1960s, they were inevitably led to comparisons with New Orleans at the turn of the century, when blues and jazz were emerging in a golden age of creativity. Were the Cunard Yanks the secret force behind the Mersey Beat and the rise of the Beatles?

CHAPTER 5

SKIFFLE

The theory that the Cunard Yanks influenced the Beatles and the explosion of beat music in Liverpool rests on the records they purportedly shared with Liverpudlians, but those who were there at the time and knew the Beatles were almost unanimous in negating the Cunard Yank influence. Bob Wooler of the Cavern Club said, "No and no again, this is another of those myths about the Cunard Yanks—I never received any records from sailors at all!" Johnny Byrne of Rory Storm and the Hurricanes concurred: "We certainly never got any material that way and I doubt that the Beatles did." Bill Harry, the publisher of the influential *Mersey Beat* music newspaper in Liverpool, also argues that the Cunard Yanks did not play an important role, asking why Southampton never developed a beat music scene despite having many more Cunard employees than Liverpool did in the 1960s.[1]

The Cunard Yanks might not have been that important in inspiring the Beatles and the 1960s beat scene, but they definitely played a vital role in bringing the musical culture of the promised land to Liverpool—only they did it well before 1960. The Cunard Yank phenomenon dates back to the beginning of the twentieth century, when Liverpool seamen brought over jazz and blues records. Immediately after World War II, they brought country and western records, and these imports did play a significant part in the development of the Liverpool music scene and in the musical education of the Beatles and their peers. Country music had been discovered by the record companies in the 1920s, and the amazingly successful Jimmie Rodgers and Carter Family recordings spread the allure of the music worldwide. Liverpool musician William Walters received a collection of Jimmie Rodgers records from his grandfather, who had lived in the Deep South during his days as a merchant seaman.[2] The American record companies considered country music a small-scale regional product until the Great Depression, when they found that rural folks were still buying records, and the rest of the nation was not repelled by the nasal twang of the vocals or the droning fiddles that sounded distinctly archaic in executive offices in New York. Country music became more popular in the 1940s, and its popularity spread to all corners of the world. During the war, soldiers from the Deep South took the music even farther afield, to the islands of the Pacific and to Liverpool, England. When American Forces (radio) Network polled their listeners at the end of the war, they were surprised to find that servicemen preferred country singer Roy Acuff—a traditionalist whose career started in the traveling medicine shows—over Frank Sinatra, the handsome young crooner who was the current star of popular music.

Many independent companies set up in the United States after the war were focused on country and its slightly more rambunctious cousin, rockabilly—a faster, louder, and more guitar-laden music that moved away from traditional topics like church and family to embrace songs about drinkin', cheatin', and driving too fast. Country music was speeding up a little, using more powerful amplifiers and cleansing itself of some unattractive hillbilly stereotypes. Performers like Hank Williams took the basic formula of country songs and turned them into upbeat, danceable pop records with delightful hooks and swinging rhythms. Unlike country in the past, which had been acoustic, the sound was electrified with the new guitars and amplifiers made by the Fender company in Southern California. Williams signed on with the new MGM label, which was formed in 1946 as the record division of the Metro-Goldwyn-Mayer film studios. Although based in Hollywood, MGM had excellent A&R men in Nashville, and they took a chance on a young singer brought to them by the influential music publisher Fred Rose. Hank Williams produced a string of hits for MGM that quickly made him an international star. Not only did records like "Jambalaya" and "Lovesick Blues" convert much of the pop audience to the exuberance and honesty of country music, they also spread the sound internationally, reaching out to converts as far away as Liverpool: "I went bananas, I went mad on it," recalled Hank Walters (who changed his name from William Walters) of that moment in 1949 when he heard "Lovesick Blues" on a jukebox in a Liverpool café. "So they got the record out, got its (identifying) number," Walters ordered it from a shop on Robson Street, and country music came to Liverpool. A Hank Williams record had the same impact on Bernie Green: "I knew as soon as I heard it that that was what I was going to do."[3] The histories of rock'n'roll recount moments of epiphany when young musicians (like John Lennon) heard an Elvis record, but there are probably just as many stories about life-changing moments upon hearing Hank Williams.

In the 1950s Walters and Green led the country music movement in Liverpool, Walters with the Dusty Road Ramblers, and Green with the Drifting Cowboys.

With groups like Cy Con and the Westernaires and Phil Brady's Ranchers, they made Liverpool the UK's center for country and western, with the most amateur and semi-pro bands as well as many clubs devoted to the music. The "Nashville of the North" also constituted the biggest audience for country records imported from America. This is where the Cunard Yanks played a significant role in bringing to Liverpool the rarer records from lesser-known and more traditional players like Lefty Frizell and Webb Pierce in the late 1940s and early 1950s.

The Cunard Yanks did bring American music to Liverpool, especially if we extend the definition of Cunard Yanks to cover all merchant seamen and steamship employees. For example, Tony Allen of the Blue Mountain Boys got his prize discs from his brother in the merchant navy. Hank Walters and John McNally (of the Searchers) told Bill Harry that they got records from seamen. Jose McLaughlin of the Timebeats, a band that played blues and jazz as well as rock, got records from his father, a merchant seaman who had opened a café near the docks. His father's wartime mates from the merchant navy knew he "was a singer who loved jazz and blues, so they brought him rare records from the USA." Charles Landsborough got records and guitars from his seafaring brothers in addition to other exotic gifts from North America, such as jeans.[4] The Cunard Yanks were especially useful in obtaining the long-playing albums that were unavailable in the UK because the English record companies usually only distributed American singles. American albums were highly prized, especially those with previously unreleased material. The serious collectors were most interested in the LPs, which they passed around and loaned out to the devoted: "One album would do for five or six streets."[5]

Many of the seamen who filled the role of Cunard Yanks came from the large African and Afro-Caribbean community on Merseyside, concentrated in the Toxteth and Liverpool 8 districts. These multicultural areas had a dynamic social life, with many clubs and entertainment venues, which was partly the result of informal segregation in Liverpool from the 1930s onward. Afro-Caribbean people joined other immigrants from all over the world in areas that had once been reserved for the Irish or Welsh. Steve Aldo remembered twelve different nationalities on his stretch of Stanhope Street. Entertainment was locally based, either in private residences, where Africans held "cellar parties," or in clubs. In the 1950s and 1960s, the West Indian Club, the Palm Grove Club, and the Pink Flamingo all had well-stocked jukeboxes of American jazz, blues, gospel, and R&B records, in addition to the latest calypso and blue beat releases.[6] The owners, employees, and customers of these clubs were drawn from the seamen who lived in Liverpool's black districts. The newly formed Afro-Caribbean groups of the 1950s—

Derry and the Seniors, the Chants—scanned the latest American releases to find new songs to perform with the same diligence as their white counterparts in the beat groups. The records from the Del-Vikings, the Miracles, and Little Richard had equal effect on Liverpool's black musicians as they did on Lennon and McCartney. But the former had the benefit of excellent relations with African American GIs from nearby bases who danced in their clubs, dated their girls, and shared their record collections with their new friends in Liverpool.

The Cunard Yanks were not the only source of rare American records in Liverpool; the huge USAF airbase in nearby Burtonwood had equal weight in bringing the sounds of rural America to Lancashire. The U.S. armed services' policy of keeping the troops entertained paid dividends for Liverpool's musicians. Many American bands were brought into Burtonwood to play, and the base had a jukebox that was the talk of Merseyside's country and blues fans. The base also employed local English bands to play their version of American country music for the servicemen, and the Drifting Cowboys played there twice a week for four years. The gigs were well paid (thirty pounds, versus the going rate of ten pounds, and all you could eat—a big attraction in hungry, rationed Britain), and they provided records so that some members of the Drifting Cowboys were able to amass large collections.

Liverpool's country groups listened to AFN to hear the latest releases and then asked their merchant seamen or American servicemen friends to obtain them from the United States. One of the consequences of stationing thousands of young men in a foreign country during wartime was the resultant number of war brides, seventy thousand in all, who added to the links with the United States. Earl Preston got records from his brother-in-law at Burtonwood. Jim Clark of the Dimensions got his American discs from his brother in the army, who had an American colleague: "He used to bring his country music LPs with him . . . so that's how I was introduced to country music."[7]

If there were no personal links to America, the record shops could always order the disc—a much easier and quicker alternative to the Cunard route! Hank Walters paid only 1 shilling, 9 pence, for his imported Hank Williams single, not a bad price for one of the first pressings made available in Europe. Large retailers like NEMS weren't the only ones who could import American records; Bernie Green got his from the Music Box, on the West Derby Road, a small shop run by a lady called Diane and her mother.[8] By 1960 several musicians in Liverpool, like Kingsize Taylor or Chris Curtis of the Searchers, owned many imported discs unobtainable in the UK. Liverpool musicians as a whole seemed to have amassed quite respectable collections of American records from retail outlets.

Drummer Paul Chiddick inherited a Dansette player and records from his older sisters in 1960. Over four years (1956–1960) they had bought discs by Buddy Holly and Elvis Presley, but also records from Little Richard, Johnny Tillotson, Eddie Cochran, and Roy Orbison—a roll call of the performers who influenced the Beatles.[9]

Outside the function of importing American records, the Cunard Yanks made important contributions as musicians and strengthened ties to the United States. Serving on board a transatlantic steamer increased the number of Liverpudlians who had experienced American cities and culture. Many, like Hank Walters' grandfather, jumped ship and lived there for long periods. From 1830 to 1930 the port of Liverpool dispatched 9 million emigrants to the United States, and most of them kept family ties with home.[10] The Beatles were members of a generation of Liverpudlians who felt especially close to the United States. They knew people who had lived in North America, friends or family members who had emigrated, such as George Harrison's sister, and if they had not toured the country as musicians, they surely would had gone there as visitors or emigrants. John Lennon's reflection that he should have been born in New York, "that's where I belong," reveals how closely he and his friends felt the pull of the American dream.[11]

Serving on a transatlantic voyage brought Liverpudlians together with American seamen and led them to exchange musical ideas. Bernie Green remembers serving on board a ship with a group of southerners and enjoying jam sessions. Seamen habitually made music to entertain themselves during the long voyages with fife, fiddle, harmonica, and especially the guitar. Some of them used the long periods of down time to learn how to play these instruments. Tommy Hicks was born in London and went to sea when he was fifteen, crisscrossing the Atlantic on the Cunard liner *Mauretania*, which made him a London-based Cunard Yank. A natural entertainer, he amused himself and the passengers on board with singing and impressions. On leave he sang and played guitar at pubs and coffee bars and sometimes as a member of a country and western group that played the small but lucrative circuit of American air force bases in England. As Tommy Steele, he was billed as the first British Elvis.

After learning how to play on board, some ambitious seamen used their sojourn in American cities to advance their careers as entertainers. Norman Milne was born in Liverpool in 1928 and joined the merchant navy during the war. During a stopover in New York, he won a talent contest in Radio City. Encouraged by this, he got a job singing in front of a dance band back in England and came to the attention of Norrie Paramor of (British) Columbia Records in 1955. He changed his name to Michael Holiday, made several hit records, and presented his own variety show on television.

THE NASHVILLE OF THE NORTH

Country music was the music the Beatles heard when they were growing up. Significantly it was remembered as the music of their childhood, not their adult years, as Ringo Starr implied to a journalist during the first American tour: "Been listening to American country songs since I was a kid, y'know. Think all of us loved the sound."[12] The country sound was an important ingredient of rock'n'roll, and its familiarity to Liverpudlians prepared them for more southern music. Although rock'n'roll was billed as a totally new sound, it was in fact the old race and country music played louder and faster and desegregated of its black and poor white signifiers. Whether it was the R&B of the former, or the rockabilly of the latter, both were moving from the periphery to the center of popular music, joined as a more accessible product called rock'n'roll. Before Elvis was the King of Rock'n'roll, he was the Nashville Cat who toured the dance halls and honkytonks of the Deep South, playing down-home country with a swinging beat. Country bands paved the way for the introduction of rock'n'roll to Liverpool, and some of them were playing the music before the term rock'n'roll was made up by radio deejays and record companies. The Liverpool music scene had its cowboys, and they dressed the part in clothes that Hank Williams would have been proud of. As John Lennon remembered, Liverpudlians, especially the Liverpool Irish, took their country music very seriously: "There's a big, heavy following of it."[13]

In Liverpool's crowded music scene, country bands played with jazz quartets and beat groups, sometimes jamming together and working through repertoires to pick up an interesting new song. There was constant intermingling of musicians and exchanges of records, guitars, and ideas. Prem Willis-Pitts swapped a copy of Chuck Berry's "Roll Over Beethoven" for George Harrison's copy of "Jambalaya."[14] The former became a standard in the Beatles' repertoire. Perhaps if this exchange had not happened, the Beatles might have ended up playing more Hank Williams songs!

The Beatles shared stages with country groups and often appeared in country music venues in Liverpool, such as the Black Cat Club. This cross-fertilization of musical styles ensured a country presence in rock band playlists. The Beatles' debt to country music can be appreciated in their early repertoire. In 1960 they were regularly playing "Blue Moon of Kentucky" (Bill Monroe), "Hey Good

Lookin'" (Hank Williams), and "I Forgot to Remember to Forget" (Elvis Presley). When they auditioned for their first radio appearance in 1962, the producer evaluated them thus: "Not as rocky as most, more country and western with a tendency to play music."[15]

Guitar-mad teenagers like the Beatles immediately noticed the twangy electric guitars of country records. The guitar stayed in the background in big bands as part of the rhythm section, and the lead voice of rock'n'roll was usually the saxophone or piano. But the electric guitar was prominent in country, and it played an important part (along with electric steel guitars and fiddles) in its signature sound. The popularity of country music in Liverpool made the electric guitar the focus of the beat groups and Merseyside an important market for these instruments in the kingdom. This was the opinion of a sales director for an electric guitar distributor, who said that Liverpool was the first city to show interest in rock'n'roll guitar playing and that "what Liverpool does, the rest of the nation follows." Liverpool musicians also believed they "identified with American guitarists more than any other city in the UK" and imagined close spiritual ties with black musicians.[16]

Not only did country and western music play its part in the Beatles' musical education, it was also a strand in another music that would have even more influence on the band. The rise of skiffle in England had a powerful effect on the musical aspirations of an entire generation of young people. Its roots were in the United States, and it was another product of the international diffusion of recordings from America to England. The year of Elvis and Davy Crockett, 1956, was also the year of Lonnie Donegan, and if you had to choose the most important single record in the development of the Beatles and Mersey Beat, it would have to be his "Rock Island Line" of that year. For those in the know, the people who actually lived through it, Lonnie Donegan and skiffle "had more influence on country music and rock music in this town than anything else." "One of the main catalysts that sparked off the Mersey music scene was skiffle music, and that was down to Lonnie Donegan." The promoter Sam Leach concluded: "No Lonnie Donegan: no Mersey Beat."[17]

TRAD JAZZ AND SKIFFLE

Skiffle came to Liverpool as part of the trad (traditional) jazz movement of the late 1940s and early 1950s. Its supporters were at the opposite end of the musical spectrum from the Teds and rockers. They were much better educated, law-abiding, and affluent, but equally devoted to music coming from the United

States. The Beatles and their fellow travelers in discovering country and R&B music in Liverpool thought of themselves as serious record collectors, but they paled against the jazz aficionados whose devotion to collecting American recordings approached the fanatical. Imagine the difficulties of obtaining recent American releases on vinyl in 1950s England, and the exertions of the Cunard Yanks in bringing them over, but then think how difficult it would be to find 78 rpm shellac records dating back to the 1920s—records that were rare in the country in which they were issued. Many of these releases were limited to small niche markets in the United States, yet some gradually found their way into the hands of UK collectors. Several shops in London, Cambridge, and Oxford sold the rare shellac discs that had originally been intended for southern jukeboxes and northern ghettos.

Some of the more dedicated collectors went back to the classic recordings from the dawning of the Jazz Age in the early 1920s, just a few years after the record companies had discovered the commercial potential of the new music called "jass." Rare recordings of King Oliver, Jelly Roll Morton, and Louis Armstrong's historic Hot Fives and Sevens formed a basic discography of the origins of jazz. Chris Barber had collected over fifty shellac discs by the time he was fifteen. With the help of a neighbor who had relatives in the United States, he obtained records by many of the lesser-known names in early blues and jazz, such as Cow Cow Davenport and Jimmy Yancey—important pioneers who were virtually unknown in their home country. English collectors delved ever deeper into African American musical history and found discs of jug bands and blues shouters dating back to the 1920s. For lack of a better descriptive term, "traditional jazz" covered all these styles.

The revival of early New Orleans jazz styles, called Dixieland, began in the United States in the 1940s, when several recordings from jazzman Bunk Johnson met with unexpected high sales. Johnson had started playing in New Orleans well before Louis Armstrong and was still going strong in his sixties when compilations of jazz and blues records were transferred onto the new long-playing albums. These re-released recordings brought a treasure chest of historic American music to the world. Trad jazz was especially popular among university students and other intellectuals, like the exis. The handful of record collectors who led the trad jazz movement in England were completely devoted to the music. The two Colyer brothers, Ken and Bill, even joined the British merchant marine so they could get across the Atlantic to the jazz clubs in New York and finally reach New Orleans—the Valhalla of all trad jazz fans. The Colyers' fascination with American culture began in the same place as it did for many other European jazz fans: the local cinema. They watched Hollywood melodramas and musicals and gradually immersed themselves in an imaginary culture of street hoodlums, gold diggers, and jazz musicians like Louis Armstrong. One of the Colyers' school friends acknowledged the importance of American films in their daydreams: "We saw them all three or four times each—streetwise kids getting Americanised very fast."[18]

The appeal of trad jazz to young Englishmen was twofold. First, it had all the marks of authenticity that gave it the cachet of the real African American experience, which was so appealing to the romantic sensibilities of young men brought up in the terraced houses or leafy suburbs of England. Second, it was relatively easy to play. Jazz was by definition virtuosic and was admired as such, but who could hope to play like Louis Armstrong or Sidney Bechet? If big band jazz was the preserve of professional musicians, trad jazz was all about being an enthusiastic amateur. You could fill up a trad jazz band with cheap substitutes for the expensive instruments mastered by the pros: washboards for rhythm, trash bin lids for drums, harmonicas and acoustic guitars for the leads, and tea chests with a pole attached to substitute for a double bass. Trad accessed jazz at its raucous beginnings and stressed the "low down" sound of the jug and string bands that brought in a broad range of influences from blues, folk, country, and even the minstrel shows. The music was hot and fast. You played with passion rather than precision: "We were a bunch of roughnecks, playing a purposely primitive music," said Ken Colyer. The musicians played not for money but for the joy of it. Author Pete Frame caught the essence of their evangelic fervor: "When a job went particularly well, they went home bristling like revolutionaries."[19] The audiences in pubs and coffee bars loved it and engaged in frenetic dancing to its fast tempos. Like the big band jazz that preceded it, trad jazz and Dixieland were dance music.

In the late 1940s Dixieland bands started to emerge in English university communities and the metropolis: the Jelly Roll Kings, Christie Brothers Stompers, George Webb's Dixielanders, and the Crane River Jazz Band. The popular music press started to take notice, and soon publications like the *Record Mirror* and *Melody Maker* were running articles on trad jazz. A few leading bands established themselves in clubs in London's West End or pubs in the suburbs. The Crane River Band (with Ken Colyer on trumpet and Bill on washboard) appeared on the BBC radio Light Programme's *Jazz Club* in 1951 and even on television, the BBC's *Picture Page*, which was the first time any British jazz band got that sort of exposure. The band played in the 1951 Festival of Britain, billed as "a Landmark in British Jazz," and a recording was issued on Parlophone.

Liverpool had its own jazz community, largely drawn from the art schools, colleges, and universities established near the city center. The Liverpool Jazz Society was a large and active showcase for the music in addition to other venues like the Storyville Jazz Club on Temple Street. The Cavern Club, "home" of the Beatles and now a major tourist destination that romanticizes Mersey Beat, actually started as a jazz club. Pictures of the Cavern a few years before the Beatles played there show an audience of serious, bearded young men with shirts, ties, and corduroy jackets.

The growing popularity of New Orleans jazz among young people in England did not meet with universal approval of the pioneers who had started the movement; they criticized the players as being mere copyists with few musical skills and absolutely no sense of the traditions. But whatever its faults, the "trad boom" opened up the joy of playing music to a lot more amateur players and started breaking down the strict lines of professionalism and class that had dominated the UK music business. It also established a number of old tunes, like "Tiger Rag," "The Saints Go Marching In," and "Midnight Special," as the trad jazz canon—the songs that everyone in the pub would know and most musicians could make a passing attempt at playing. These standards became the repertoire of trad jazz and its raucous and informal offspring, "breakdowns" and skiffle.

The English record industry had little interest in trad jazz. The companies fixed their attention on the established stars and the latest photogenic American crooner. They knew that the trad jazz audience was small and too discerning, and the players were traditionalists who did not yearn to make hit records. But one song, one charismatic performer, and one phenomenally successful record would change all that, just as it had done many times before in popular music.

Anthony Donegan became fascinated with jazz by listening to it on a BBC radio program called *Radio Rhythm Club*, which was broadcast weekly during the war. Donegan did his national service in Austria, where he spent much time in the American sector, listening to AFN and immersing himself in the music of African Americans and poor country whites. This was his inspiration. Back in England he changed his first name to Lonnie in honor of the blues legend Lonnie Johnson, bought a cheap guitar, and started playing the trad jazz clubs in London. He listened to blues and country greats on the Library of Congress records he obtained from the American Embassy and checked out the secondhand records in London's left-wing bookshops, which often stocked the "real" music of American workers. In this way he absorbed the art of Jimmie Rodgers, Big Bill Broonzy, and Huddie "Lead Belly" Ledbetter from their recordings. He was ambitious as well as talented, and he soon got jobs playing guitar and banjo in the bands led by Colyer and Barber. In a break between sets, Donegan entertained the crowd by playing old American folk and blues songs on an acoustic guitar. His performance style owed a lot to the British music hall tradition and a long line of comedians who used music in their acts. He did a lot of comedy songs, making faces and mimicking American voices. These short informal interludes became quite popular, and before long they were a permanent part of Chris Barber's show. Bill Colyer named them "skiffle," after an obscure Dixieland jazz band, and the name stuck.

In 1954 the Barber band went into Decca's studios to put together an LP entitled *New Orleans Joys*, which was designed to test the market for trad jazz. They allegedly added Donegan's version of Lead Belly's "Rock Island Line" as an afterthought and over the objections of the producer. The master tapes languished in the studios for some time, but once Decca released them, the unexpectedly high sales persuaded studio executives to market some of the tracks as singles. Decca released "Rock Island Line" in 1955 as a comedy record. Its spoken introduction led gradually into the song, which soon speeded up to a frantic tempo, and Donegan's nasally, often inaudible vocals set it apart from other trad jazz records. It got radio play as a novelty record, always popular with British listeners who were accustomed to laughing along to their favorite comedians, but as more and more requests came into the BBC, "Rock Island Line" entered the pop charts. Sales increased rapidly and it sped into the British Top 10 in January 1956—a chart dominated by Bill Haley's "Rock Around the Clock." It stayed in the Top 10 for several months and sold 3 million units. Then the record did the unimaginable for a British act. Released on the London label in the United States, it broke the *Billboard* Top 10 in March 1956. Lonnie Donegan got to tour the United States, playing gigs from the Midwest to the Deep South, appearing on radio and on television in the nationally broadcast *Perry Como Show*, where he sang and played a sketch with an American actor called Ronald Reagan.

"Rock Island Line" began its life deep in the American South as part of the folk traditions handed down from generation to generation. Lead Belly could have learned the song during his time in prison or picked it up on his tour of southern penitentiaries with the Library of Congress ethnomusicologist John Lomax in 1933 or 1934. It says a lot for the potency of the record as a preserver of culture that the song survived the Depression and all the trials and tribulations that a poor, uneducated black man had to face in 1930s America. That it was to have such an impact on the British music scene shows the power of the record as a diffuser of popular culture across national boundaries.

"Rock Island Line" began the skiffle boom in England. There was a flurry of similar recordings, a host of new bands and venues for them to perform, and a run on washboards. The familiar country sound and origins made it extremely popular in Liverpool, and there might have been as many as five thousand skiffle bands in the UK as a whole.[20] The skiffle boom broke the ground for the beat boom a few years later. Amateur and often underaged skiffle groups were precursors of the beat groups, and they established the basic formula for the garage bands that would dominate rock'n'roll for decades after. One skiffle group, formed in Liverpool and called the Quarry Men, would have special significance.

Skiffle was based in the distant American past, in the mythical Deep South and West, just like the Davy Crockett stories. It was fundamentally nostalgic for a bygone time and music, but it also had a modern appeal to the young. The generation of teenagers born after the war did not have time to learn an instrument; what they needed was immediate gratification. The trad jazz players born before the war had picked up trumpets and cornets like their idols Louis Armstrong and King Oliver and then spent years mastering them, along with other difficult instruments like clarinets and trombones. The skiffle generation did not have that much patience or perseverance. Paul McCartney liked the idea of playing a trumpet but was turned off when he found out it required building up a callous on his lips. Quarry Men guitarists Eric Griffiths and John Lennon tried taking lessons from a guitar teacher: "We only went twice, because the chap wanted to teach guitar properly, whereas we wanted instant music."[21] Skiffle was instant music played with cheap acoustic guitars, which were readily available for as little as five pounds in one of Liverpool's numerous secondhand shops. Guitars were easy to play and carry, and this was important. Because private automobiles were rare in 1950s England, amateur bands went to their gigs on public transport. The Beatles were riding buses to their engagements as late as 1961. Pity the drummer and the stand-up bass player.

The skiffle music performed by hundreds of amateur bands was crude, often rude, but pulsating with noisy excitement and enthusiasm. It was solely a British phenomenon, hardly attracting any attention in the land where the music was born. Skiffle was fun, and anyone could do it by mastering a few chords on an acoustic guitar. It marked the beginning of the involvement of thousands of young men in a business that had been the preserve of highly trained and unionized adults. Skiffle was the force that pulled John Lennon and Paul McCartney out of the audience and onto the stage. It made them think of music as something you *did*.

"Rock Island Line" played a seminal role in the history of the Quarry Men. John bought the 78 rpm shellac disc in 1956. He then convinced his classmate

Pete Shotton to form a skiffle band, named the Quarry Men after the school they attended. The Donegan classic was the first song in their limited repertoire, and they went on to master all the skiffle standards that Donegan popularized, including "Midnight Special" and "Worried Man Blues" (the latter remained in the Beatles' repertoire until at least 1960). George also heard the record and took note of Donegan's guitar. He said later, "Lonnie and skiffle seemed made for me . . . it was easy music to play if you knew two or three chords, and you'd have a tea chest as bass and a washboard and you were on your way."[22] Paul McCartney was in the audience when Donegan appeared at the Liverpool Pavilion, and Donegan's shows at the Empire Theatre inspired both Harrison and McCartney, leading both to ask their parents for guitars, according to Bill Harry. George was so enamored that he managed to get the star's autograph.

"Rock Island Line" and the skiffle craze took over the existing channels of finding and developing musical talent. There were skiffle contests, talent searches, and guitar lessons in pubs, coffee bars, youth centers, dance halls, and cinemas all over the country. Skiffle took over that bastion of cheap variety entertainment, the talent show. Local, regional, and national contests sought out amateur bands to appear in skiffle auditions organized in every variety playhouse, church hall, and cinema in England. Skiffle brought the Quarry Men onto the stages of some of Liverpool's great theaters when the band auditioned at talent shows at the Empire and the Pavilion. They competed with other groups, such as the Dusty Road Ramblers, formed by Eddie Clayton in 1957, which included his next-door neighbor, Richard Starkey (later known as Ringo Starr).[23] The Quarry Men even played before the snooty jazz crowd at the Cavern Club, a gig that did not go down well and finished any hopes that the band would ever appear there again.

Skiffle was a run-through for record producers and artist managers that served them well when an even bigger musical fad energized English record buyers. Agents and record company A&R men cruised the coffee bars and skiffle clubs of London, ready (as the musicians imagined) to offer a recording contract on the basis of a performance or a chance meeting. One of them was George Martin, who was searching for trad bands and skiffle players for the Parlophone label he had just been assigned by EMI. He visited the 2i's coffee bar and signed up a trad band called the Vipers in November 1956, but unfortunately he missed their singer Tommy Steele.[24] The story of Steele's discovery provided the inspiration for an army of amateur players and for the retelling of the pop music rags-to-riches story. The stage play *Expresso Bongo* followed the rise of a kid playing an acoustic guitar in a Soho coffee bar who was catapulted to fame by a grasping and devious manager. It wasn't too far from the truth.

Most teenagers usually discovered skiffle and trad jazz in coffee bars. The 2i's in Soho was recognized as the birthplace of British rock'n'roll, but all over the country, entrepreneurs were buying Italian espresso machines, tables, chairs, and those ubiquitous red checkered tablecloths and opening up for business. In Liverpool Allan Williams and Mona Best saw the opportunities and gave amateur groups like the Quarry Men their first chance. They built clubs and coffee bars in their basements and turned their homes into stages for teenage bands. They might not have had many amenities, but they had cheap food and drink and lots of atmosphere. Cynthia Lennon compared Williams's Jacaranda Club to Dante's *Inferno*: "The heat, sweat and noise nearly knocked you over as you struggled and shoved to descend the narrow heaving stairs."[25]

Skiffle bridged the country and western past with the rock'n'roll future. You could play skiffle at a church social or trad jazz in a pub, but you could fill up a large hall with people who wanted to rock. The increasing audience size was the critical ingredient in the rise of rock'n'roll, and the venues established for skiffle and trad jazz followed the audience into rock, becoming "jive hives" or beat clubs. Even the elitist Cavern Club finally dropped its resistance to rock, and its beautiful friendship with the Beatles began.

The flourishing Liverpool beat scene reflected the growth of a paying audience as much as it expressed the talent of local musicians. By the 1960s the town center was packed with bars, clubs, and meeting places. It was teeming with nightlife. The music halls and pubs that dated back to the Victorian era were joined by a mass of new venues, jazz, country, and folk clubs, as Liverpool's entertainment district moved out of the city center and expanded toward the suburbs. The Merseyside Clubs Association had at least three hundred working men's clubs, which served alcohol and provided entertainment. Numerous factories, companies, and political associations held regular socials and dances, as did churches and synagogues. One Liverpudlian noted, "Every church hall, ballroom, town hall, and skating rink where a stage could be erected and admission charged was holding a weekend dance."[26]

Skiffle gradually undermined the hold of jazz on Liverpool audiences and promoters. It brought in a younger and more energetic crowd that was far better for business than jazz aficionados, who just sat around listening and did not consume much food or drink. The audience for youth music was expanding so rapidly by the end of the 1950s that many skiffle bands made the transition into rock simply because many more people wanted to listen to it. This was the path taken by the Quarry Men as they evolved into the Silver Beetles, and it was the same for many of the other leading Mersey Beat bands: the James Boys became

Kingsize Taylor and the Dominoes, the Raving Texans evolved into Rory Storm and the Hurricanes (with Ringo on drums), and the trad group the Blue Genes became the Swinging Blue Jeans. The Searchers started as a Liverpool skiffle combo, as did Gerry Marsden and the Pacemakers. Skiffle prepared the way for rock'n'roll because it engaged an army of amateur musicians and brought them together with a youth audience that became a mass market for entertainment in the United Kingdom.

CHAPTER 6

ROCK'N'ROLL COMES TO BRITAIN

Skiffle might have gotten each of the Beatles on stage, but rock'n'roll took over their lives and pushed them into becoming professional musicians. John Lennon was not alone when he said that rock'n'roll was like a religion to him. It encompassed a look, a lifestyle, and an attitude as well as music. While skiffle and trad jazz were played by amateurs having fun, rock'n'roll was a business, a dynamic and profitable business that brought significant changes to the British pop music industry. It also transformed the lives of the musicians who made up the Beatles.

Reading the histories of rock'n'roll today, you might be surprised that it got off the ground in England at all. Although the UK had a few small independent record companies, no Sun or Chess existed to discover the talent and start the movement with some historic records. Most of the new record labels formed in the UK after the war were subsidiaries of multinationals in electrical manufacturing (such as Pye and Philips) or of film studios that were eager to exploit the profitable record business (such as Top Rank).[1] These large diversified businesses saw opportunities in the postwar popular music market (just as their counterparts in the United States did), but preferred to be followers, rather than leaders, in the introduction of new music. The Oriole company was the only independent that might have made a difference in the UK. Formed by the Levy Company in the 1920s, Oriole was revived in 1950. It was a big operation with two pressing plants, but its business was mainly in cover records, which were sold in Woolworth's stores. Oriole became an important presence in the Liverpool music scene under the direction of A&R man John Shroeder. He signed local singer-songwriter Russ Hamilton from a skiffle group in 1957 and pushed Hamilton's "We Will Make Love" to Number 2 on the English pop charts.[2] But this was a romantic ballad that was not much different from all the other easy-listening records, and perhaps its success kept Shroeder from recording more innovative

Liverpool music. He did not see the potential of the Afro-Caribbean groups who might have come up with a musical hybrid similar to rock'n'roll.

The spread of new music in England was hampered by the state-owned British Broadcasting Corporation (BBC), which acted as a conservative gatekeeper to the airwaves. Unlike the highly competitive American radio industry, broadcasting in the UK was a monopoly. The war years strengthened the BBC's hold on broadcasting and established the ruling oligarchy of EMI and Decca in the record business. All of them dismissed rock music as another American fad that would soon go away like the other short-lived musical novelties of mambo and calypso. The industry professionals on both sides of the Atlantic were often biased in their assessment of youth music, condemning it as the work of amateurs with low musical standards and poor recording facilities, which was often the case. They perceived the youth market as fickle and unsophisticated, and they continually underestimated its size. There was almost a professional pride in ignoring the youth audience in the offices of the BBC. Disc jockey and television presenter Jack Payne summed up this attitude in an article he wrote for the *Melody Maker*: "Should We Surrender to the Teenagers?" The powerful A&R man Mitch Miller was asking the same question in the United States. The answer was always no.

The BBC had often discriminated against popular music on aesthetic or moral grounds, and because it had the monopoly on broadcasting, its refusal to play a song was usually the end of it. Yet Bill Haley's "Shake, Rattle and Roll" was never played by the BBC and still managed to slip into the NME chart at Number 13 in December 1954, and by January the next year it was in the Top 5. How could anyone have heard this record before they bought it? The answer could be found in broadcasting media on the periphery of British entertainment, the jukeboxes found in select coffee bars (like the 2i's) and the record shops that catered to the growing number of rock'n'roll cognoscenti. These few outlets for new music managed to circumvent the oligarchy that controlled the marketing of records in the UK and brought the sound of rock'n'roll to adventurous teenagers.

American armed forces radio still broadcast in Europe in the 1950s and 1960s, and it was popular with listeners. The English newspapers published its schedules along with the BBC's Light, Home and Third programmes. It was joined by Radio Luxembourg, a commercial station established in the grand duchy by a French company in 1933. During the war it became part of the American propaganda effort, and the U.S. Armed Service planned to make it the "Voice of America" in Europe in the postwar years, but commerce prevailed over propaganda, and the British-language service of Radio Luxembourg gained strength in the 1950s as an important new vehicle for diffusing American music.

Established at 208 meters on the medium wave, the station broadcast pre-recorded commercial programs produced by European and American record companies, who sent their latest releases to Luxembourg, where the biggest radio transmitter in the world spread them far and wide. You could only pick up Radio Luxembourg's signal well after dark in England, and it was a struggle to keep the dial fixed on 208 and the signal coming in. Yet thousands of English teenagers did it (including all the Beatles and me), and it became a ritual to listen to the popular music segments broadcast on Saturday and Sunday nights. The highlight was the two-hour *Jamboree* that started at 8 p.m. on Sundays, an "exciting, non stop, action packed" program that starred thirty minutes of deejay Alan Freed babbling from New York City and playing records that we never dreamed existed. It was Radio Luxembourg that brought Elvis's "Heartbreak Hotel" to the attention of four young men living in Liverpool in the 1950s.

Radio Luxembourg dramatically increased the profile of rock'n'roll in Europe, and there cannot have been many English schoolboys who had not heard some of it by the end of the 1950s. Soon the British record companies were digesting the sales numbers for rock records in the United States and their distribution over several different audiences—Elvis scored hits in the American pop, country, and R&B charts simultaneously—and thinking that this might not be a short-lived fad. By this time Elvis Presley was no longer the Nashville Cat representing a tiny record label in Memphis but the King of Rock'n'roll in the pay of RCA-Victor, a major presence in the American music industry. Smaller independents like Sun and Atlantic had led the way into rock'n'roll, but as soon as the big companies realized the potential of the new music, they overtook the independents and their African American stars with more palatable white singers.[3] RCA had taken a big gamble on Elvis, paying a record fee to get his contract from Sun and hoping to turn a regional attraction into a national figure. This it accomplished with help from NBC radio, CBS television, and movie studios Paramount and MGM. The music and image might have seemed rebellious, but Elvis Presley the film star and celebrity was the perfect employee of the Empires of Sound.

These multinational business organizations imported rock'n'roll into the UK. It crossed the Atlantic in the same way as all the other American entertainments that had delighted English audiences, and once established, it was copied by European musicians and record companies in a practice that went all the way back to the minstrel show. As soon as the English entertainment industry realized that rock'n'roll appealed to a mass market, they began to make copies of American originals. This practice of "covering" was especially strong in Great Britain and reflected the special conditions of broadcast entertainment. The BBC had a complete monopoly of Britain's radio waves and resisted the move into programming based on disc jockeys playing records—the "needle time" that was the mainstay of American radio broadcasting after the war. In fact, the influential rock deejays who did so much to spread the music in the United States were entirely absent in Great Britain. British radio preferred to reproduce popular records with its own in-house artists, cutting down royalty payments and keeping British musicians fully employed, a strategy supported (and sometimes enforced) by the British Musicians' Union. Their American counterparts had also recognized the threat of pre-recorded musical programming after the war and tried to ban it using strikes in the 1940s, but to no avail. The rise of the disc jockey in the United States, a professional broadcaster of recorded music, played an important part in the introduction of rock'n'roll.

The first English rock records were covers made by pop singers or large dance orchestras. Bill Haley's 1953 "Crazy Man, Crazy" was covered by a popular singer from Liverpool called Lita Roza, who also produced copies of R&B songs originally performed by Ruth Brown and Ella Mae Morse. Roza was the vocalist for the Ted Heath Band, which had mined the American R&B charts for songs that might appeal to the stodgy, elderly listeners of the BBC's Light Programme. They even found a Sun record written, sung, and recorded by Alabama country singer Hardrock Gunter about his hometown. "Birmingham Bounce" was a local hit in Alabama and Tennessee that had caught the attention of Sun Records, who purchased Gunter's master and re-released the song on its label. In England you could purchase the Ted Heath version on a 78 rpm disc!

The BBC liked to talk about "harmonizing" American songs to fit the particular musical tastes of the British public, which in the minds of rock fans meant watering the excitement down and slowing the tempo to a leisurely pace. It rearranged two classics of early rock'n'roll from Bill Haley for the sedate tempos of the dance orchestras that dominated English radio: "Shake, Rattle and Roll" was reinterpreted by Jack Parnell and his orchestra, and "Rock Around the Clock" was mutilated by the Big Ben Accordion Band.

The English record companies scanned the advance copies of American records in search of hits. The innovations of small "unbreakable" 45 rpm vinyl discs and the transatlantic jet service facilitated the rapid diffusion of American records. Such was the speed of the transportation and distribution that the cover might appear in the same chart as the original. In 1956 American pop crooner Guy Mitchell covered a Marty Robbins country song, called "Singing the Blues," for Columbia Records under the guidance of Mitch Miller. It topped the U.S. charts in November and entered the British Top 20 on the European Philips label on December 8 that same year. One week later Tommy Steele's version challenged Mitchell's song in the English charts, and throughout that month his Decca record followed Mitchell's to the top. During January they swapped the Number 1 and 2 positions and stayed in the charts until February. Two years later Michael Holiday quickly took another Marty Robbins song to the top of the British charts, beating out other covers by English entertainers Alma Cogan, Dave King, and Gary Miller.

An alternative to covering American songs with in-house talent was the more expensive licensing agreement that allowed a British record company to release an American recording on its own label. EMI and Decca were international operations, and thus identifying and obtaining foreign master recordings was part and parcel of their daily business. They kept an especially close watch on American teen music because it constituted the major part of their own popular music catalogs. EMI had licensing agreements with the two major players in the United States—RCA-Victor and Columbia—and records produced by them accounted for more than half of EMI's sales in the early 1950s. As a multinational company EMI had recognized the importance of American music in the global entertainment business since the 1930s, yet its conservative attitude to new technology would hinder its acquisition of American product. The Empires of Sound were linked together, but each part could exercise the option not to adopt the new machines or music coming from the other side of the Atlantic. The management of EMI was not ready to accept the new vinyl records developed by Columbia and RCA, and this was the reason these American companies broke the licensing link. After losing these licensing deals, EMI purchased a controlling interest in America's third largest company—Capitol—in 1955 to get access to the American recordings that filled up its pop music catalog.[4]

The English majors were licensing so much music from America that they set up special labels to market it in Europe. British Decca had formed London Records to sell its recordings stateside because the split with U.S. Decca in the 1940s did not allow it to trade under its name in the United States. Decca then used the London label in the UK to distribute selected recordings licensed from American independents like Chess, Dot, Atlantic, Sun, and Specialty, including some R&B masterpieces like "Tutti Frutti" that would have such an impression on the Beatles. Decca pressed the discs in England but added the imprint "American Recordings" beneath the London title on the label. British Decca also managed the Coral label, which brought Britons the music of Buddy Holly and the Crickets. EMI created its Stateside label to compete with Decca's London discs in the UK, issuing such important records as the Isley Brothers' "Twist and Shout."

As rock'n'roll grew more popular in England, the majors moved from licensing records from the giants like RCA to approaching the smaller American independents that had discovered the R&B talent. Ironically, EMI's Stateside made licensing agreements with the American independents Vee Jay and Swan—the first American companies to release Beatles recordings. The licensing process worked both ways. The Searchers followed the success of the Beatles in the English charts, and their hits on the Pye label were licensed for American release on Mercury and Liberty discs.

If discriminating English record buyers were repulsed by half-hearted covers of American hits from local bands, they could soon buy the American original from a British company who had licensed it. For example, Bill Haley's recordings appeared on Brunswick in the UK—the label American Decca used to market its recordings overseas. Even the rare and exotic R&B records that could claim to be the forerunners of rock'n'roll were gradually made available in their original form to English record collectors. The Chords' "Sh-Boom" of 1954 was one of the first R&B records to get the attention of white America. It appeared on the B side of a single on the Atlantic label and was quickly covered by several American, Canadian, and English groups, but it was also licensed from Atlantic by EMI and appeared on its Columbia label in England.

The British record industry quickly mastered the techniques of processing American songs into English records or acquiring the licensing rights and releasing the original. The hit songs of R&B performers like Little Richard took some time to work their way through American companies' A&R organizations, licensing agreements, and British distribution deals. It could take as long as eighteen months to bring a Little Richard or Chuck Berry song to English listeners in the mid-1950s, but by 1957 the process had been streamlined. Buddy Holly's "That'll Be the Day" equaled Elvis Presley's "Heartbreak Hotel" in the effect it had on budding British musicians. Each of the Beatles took notice of this record when it appeared in English record shops only about a month after its American release in August 1957.

Not only was "That'll Be the Day" a great song, it had some spectacular guitar playing that was not too difficult to copy. Holly's basic country licks did not have the cool articulation of Chet Atkins', but he played with so much confidence and volume that his guitar breaks simply jumped out of the disc. He covered both lead and rhythm guitar, and he used nice, easy chords effectively. Consequently, "That'll Be the Day" was quickly inserted into the repertoire of every skiffle band, and the Quarry Men were so pleased with their version that they chose it for their first recording, as did many other Liverpool skiffle groups. Holly's records became an instruction manual for a generation of English rock guitarists, including George Harrison, who borrowed them from a friend so he could copy them, and Brian Rankin (Hank Marvin of the Shadows), who impressed record producer Mickie Most: "He had all the Buddy Holly licks off before anyone else could play them."[5] Buddy Holly was an ordinary-looking young man with a big smile and thick, black-framed glasses, but his guitar playing and songwriting talents impressed the amateurs who listened to him. John Lennon said that hearing what Buddy Holly could do with three chords made a songwriter out of him. True to form, a British company quickly produced a cover version of "That'll Be the Day." EMI picked a young singer who worked in their factory, and Larry Page actually got to pack his own single, described by Pete Frame as "among the most excruciating records ever made."[6]

The British record companies had mastered licensing by 1960, when a homegrown cover could be quickly eclipsed by the speedy release of a licensed original. "Will You Love Me Tomorrow" was written by composers Carole King and Gerry Goffin, who made the Brill Building in New York City rock'n'roll's Tin Pan Alley. It was recorded by a vocal group called the Shirelles, who formed in Passaic, New Jersey, and were discovered by local entrepreneur Florence Greenberg and signed to her small in-house Scepter label. Greenberg often sold promising Shirelles masters to American Decca, who marketed them nationally and started a trend of all-female singing groups that would have tremendous implications for African American pop music (think Tamla-Motown) and an influence on the music of the Beatles. "Will You Love Me Tomorrow" sold very well in the United States, and at the same time that a copy was being circulated around the corridors of British Decca as suitable material for a cover, the newly formed Top Rank label acquired the license to issue the record in the UK and quickly put it out in January 1960. Decca gave the song to Mike Berry, one of its rock'n'roll starlets, but Top Rank's Shirelles original crushed it in the charts. The Beatles must have got hold of the Top Rank single because they gave the B side, "Boys," to their drummers to sing, and Ringo Starr's version appeared on their first album. Passaic, New Jersey, to Liverpool, England, is a long way, but the international record industry was making the connection easier and easier as it mined the mother lode of rock'n'roll.

Both big and small companies became involved in the international movement of records as rock'n'roll grew more popular, and small English independent labels made licensing deals with their American counterparts. Joe Meek's record company was based in his home studio at his London apartment, but he still managed to distribute his songs in the United States through licensing arrangements with independents like Dot Records.[7] Oriole obtained the rights to release the first Motown records in the UK, including songs by the Contours ("Do You Love Me") and the Miracles ("You Really Got a Hold on Me"), which were Mersey Beat standards. (Later the rights were acquired by EMI and issued on Stateside.) Oriole's hit record by Russ Hamilton was picked up by the American independent Kapp in 1957, but the A and B sides were mistakenly switched, and instead of "We Will Make Love," the B side "Rainbow" broke the American Top 5. Hamilton later recorded with MGM in Nashville.[8]

The coming of rock'n'roll to Liverpool wasn't accidental, or lucky, or due to a special geographic and commercial situation. It was part of the operations of the Empires of Sound. Their experience in exploiting new music and seamlessly incorporating it into their networks of promotion and distribution brought the sound of rock'n'roll and rhythm and blues to England just as it had delivered swing, trad jazz, and country. Yet the Cunard Yank mythology was so attractive that it refused to go away. John Lennon embraced the myth when he was interviewed in the 1970s. He said that Liverpool was a cosmopolitan city: "It's where the sailors would come home with the blues records from America on the ships." Pete Frame could not resist the story of an Elvis fan that went to a London record store whose owner had made a profitable arrangement with a merchant seaman to bring in American records. The customer wanted a copy of Elvis's "Hound Dog," and once it was located, the store manager asked if he was interested in hearing the original, as performed by Big Mama Thornton on the independent Peacock label based in Houston, Texas.[9]

Yet by the end of the 1950s, there were enough formal links between the English and American record producers to make such exertions, and so many good stories, unnecessary. In 1962 a correspondent of *Mersey Beat* complained about the difficulty of obtaining R&B and blues records in Liverpool and wrote about a flourishing black market based on "cherished" personal collections of American discs. A spokesman from a record wholesaler immediately corrected him by pointing out that the records he discussed were available for purchase at NEMs in Liverpool and at record shops in Birmingham, Coventry, Manchester, Sheffield, Cambridge, and Bournemouth! The spokesman added that his company distributed a wide range of blues and R&B in the UK, including records from Little Walter, Howlin' Wolf, and Bo Diddley.[10] There was never any need for Cunard Yanks to bring precious American R&B and rock'n'roll records to eager teenagers in Liverpool, for the Empires of Sound were already doing it for them.

AN ENGLISH ELVIS

With rock'n'roll so popular with English youth, the search for a homegrown Elvis, an English rock'n'roller who could make the music his own (and keep the profits in the UK), took on some importance. The men who took the lead in creating truly English rock'n'roll came from backgrounds in artist and theatrical management, record retailing, and independent studios. They combined their knowledge of distributing records with their promotional sense to manufacture the first wave of English rock'n'roll stars, starting with Joe Meek, whose "Telstar" by the Tornadoes went to Number 1 on the *Billboard* charts. He produced a string of rock and pop records with vocalists who had faultless good looks (like John Leyton), lukewarm covers of American hits, and lots of special effects in the background. In his memoirs Meek stressed the importance of making records for a teenage market, and he was well ahead of all the British record companies in this respect. The most important rock Svengali was Larry Parnes, who came from a Jewish family background in the clothing business and pursued a career in the theater. Parnes had the connections with the venues (including the all-powerful Delfont booking organization) and record companies but was not that interested in the music. Some of his signings were completed before even listening to the performer. For Parnes it was the look—the gold lamé suits, the mass of Brylcreemed hair, and the photographs he could place in the tabloids—that was more important than the songs he bought from Tin Pan Alley writers. Parnes was a showman with theatrical flair rather than a manager committed to the music.

Parnes groomed Tommy Steele to be the first British Elvis, organizing a record contract from Decca, a tour of provincial music halls, and some well-choreographed press stories. Although billed as the country's answer to Elvis Presley, Steele's career took much the same course as the other successful variety entertainers he was destined to follow: extensive touring around the country, radio and television appearances, a turn on the *Royal Variety Performance* at the London Palladium before the royal family, and finally films. Steele's Elvis-like ascent attracted the media, and he also bore the brunt of growing fan hysteria. The screaming at his shows was quickly noted in the press, and he narrowly escaped injury when a concert in Scotland got out of hand.

Although Parnes was based in London, he "discovered" the first generation of Liverpool rock'n'rollers. Ronald Wycherley used the same Liverpool recording service to make his own demo that the Quarry Men employed for their first record. He sent this to Parnes along with his photograph (very important) and was signed to a contract with Decca. As Billy Fury, Wycherley enjoyed a successful career in the early 1960s with a string of hits. Another Liverpudlian singer, John Askew, also wrote to Parnes and was signed to Philips Records. He was given the name Johnny Gentle. Parnes dominated the early years of rock'n'roll in England with a stable of handsome young men who wore the suits he chose for them, sang the American songs he found for them, and adopted the names he coined for them.

Larry Parnes' stable of attractive young singers followed the well-established tradition of copying Americans. They joined a long line of English entertainers who built their careers on copying African American material and sounding American. It started with those music hall performers, like Jack Lennon, who took songs and dances from the minstrel shows that toured Europe in the late nineteenth century. After nearly seventy years of imitating black Americans, this tradition was so strong in England that skiffle players like Lonnie Donegan could offer up weak imitations of American southern accents with no hint of embarrassment or self-consciousness. (Donegan had copied Lead Belly's version of the song exactly.) Similarly, Liverpool's country bands did impressions of American accents. Hank Walters reported that when the American servicemen heard the Drifting Cowboys, they "couldn't believe it cos we sang American style."[11]

When it came time for the Beatles to audition for Larry Parnes, they did their best to look and sound American too. They had tried to engage the nascent British rock industry at ground level, a year before Epstein turned up at the Cavern Club to take a look at them. Larry Parnes visited Liverpool to engage backing bands for a package show starring American rockers Eddie Cochran and Gene Vincent. The Quarry Men were still struggling along without a regular drummer and with very few gigs, but Parnes allowed them to audition. On the big day he turned up with his protégé Fury and announced that one lucky band would get to back up Fury on tour. Alas, the Quarry Men did not get the Billy Fury gig, but the group was overwhelmed at winning a consolation prize—backing Johnny Gentle and Duffy Power for ten exhausting, freezing nights in Scotland. They changed their name and took the first step toward becoming professional musicians. The Quarry Men had covered several bases of imported American music, country and western along with skiffle, but the Silver Beetles was now a rock'n'roll band.

The search for a suitable English Elvis continued while the Silver Beetles served their apprenticeship. An ex-variety player called George Ganjou saw a skiffle band called the Drifters, and although he had no previous experience in artist management, he was so impressed with the way the girls reacted that he took the band on. They were led by Harry Webb, a good-looking young man who could do perfect imitations of Elvis. Webb's rock'n'roll epiphany came when he saw

Bill Haley play the Edmonton Regal cinema in London in 1957; he was inspired to form a band and make a demo at the HMV record store in London. Norrie Paramor, the recording manager of EMI's Columbia label, signed the group and gave it a bland American record to cover, "Schoolboy Crush," by Bobby Helms. On the B side of their single was a song written by Ian Samwell, a guitarist for the Drifters. Paramor used his influence to get them on Jack Good's *Oh Boy!* television program, which was an important showcase for pop music. Good did not think much of "Schoolboy Crush" and insisted the Drifters play the B side, "Move It." They did, and this song caused a sensation in British rock'n'roll circles when it was released in 1958. Considered by many music critics and musicians (including John Lennon) to be the first British rock song, "Move It" sounded so authentically American that it came as a surprise to many listeners that it was performed by an English band. Ian Samwell admitted that the Drifters copied the American masters of rock and R&B but argued that they did it better than the rest of the English groups: "We were closer to the real thing than anyone else around."[12] That is to say, their copy was considered the most accurate.

The Drifters had to change their name because the American group of the same name objected to it and threatened legal action if the band released records in the United States—an indication of the importance of the American market to British rock and pop acts. So the band became the Shadows, Harry Webb became Cliff Richard, and his guitarist, Bruce Rankin, now went by the name Hank Marvin—an American name for an English guitarist playing American music with an American guitar. Hank Marvin wore those thick glasses that everyone associated with Buddy Holly and took great lengths to acquire the same Fender Stratocaster that his hero played. Although everyone thought of the Shadows as the pioneer English rock band, the sound was all American. Marvin's Stratocaster produced bright upper registers that were recognized as the "Fender sound," shorthand for a clean, trebly sound with stinging highs. Rock'n'roll guitar was usually manipulated by reverb—the echoing, ominous sound that had made the early Elvis records so distinctive. Hank Marvin used an Italian echo device, the Meazzi, and cranked out heavy vibrato with the tremolo arm of his Stratocaster.

Cliff Richard and the Shadows started the trend of having the singer fronting the instrumentalists, in the three-electric-guitars-and-drums lineup that became the standard for rock groups. Liverpool was full of them, with names like Gerry and the Pacemakers and Billy J. Kramer and the Dakotas. Cliff Richard became the most successful and long-lived English Elvis, but it was the Shadows that had the most influence on the Beatles. They were the first important guitar band in England that did not need a handsome singer up front to have a hit record. Their guitar instrumental "Apache" was a Number 1 hit in 1960 and set the stage for the British adoption of the rock instrumental genre that made American guitarists like Duane Eddy and Link Wray stars. The Shadows followed "Apache" with a string of instrumentals all constructed the same way, a solid repetitive beat produced by an electric bass, with a reverb-laden lead guitar picking out the notes of the simple melody. In the first four years of the sixties, they produced three or four instrumental records each year, and most of them made the Top 5. Even the names of their hits reflected the English admiration for American popular culture: "Apache" was inspired by the 1954 Western of the same name, and it was followed by "FBI," "The Frightened City," "Shindig," and "Mustang."

The Shadows were extremely important in the development of the Beatles' music. The first original song the Beatles ever recorded (for Polydor in Germany in 1962) was a Shadows-like instrumental appropriately entitled "Cry for a Shadow." Much of the Silver Beetles' repertoire was instrumentals, and the people who listened or auditioned them noticed the Shadows' influence.[13] Although Cliff Richard and the Shadows are virtually unknown in the United States today, they dominated British pop music at a critical time. In 1963 George Harrison said that if the Beatles did as well as Cliff and the Shadows, "we won't be moaning." Brian Epstein is famous for his belief that the Beatles would be "bigger than Elvis," but when he wrote to the record companies in 1961, he claimed that his band had written a song as good as a Cliff Richard and the Shadows hit.[14] In his liner notes for the album *Please Please Me*, Tony Barrow quotes the BBC's Brian Matthew's tribute, that the Beatles were "visually and musically the most exciting and accomplished group to emerge since the Shadows."

Record companies in North America had scant knowledge of the British rock scene except for Cliff Richard and Shadows. They were the standard against which the Beatles were to be measured. But the Shadows were definitely not cool; their bass player, Jet Harris, was cool—he had dramatic good looks, a blond quiff, and an impressive Fender Precision bass (the first in England)—but after he left, the band as a unit was not cool at all. There was no hint of rock'n'roll rebelliousness in their music or in their stage show, where they made the same choreographed steps—the "Shadows walk"—which was copied by an infinite number of amateur bands. For all his pouting and Elvis moves, Cliff Richard was safe and predictable, and the members of the Shadows were bland and silent.[15] This was the norm in the British popular music scene when the Beatles burst into the spotlight.

BREAKTHROUGH

Historians of the Beatles and many of their fans point to a concert at Litherland Town Hall on December 27, 1960, as a turning point in their career. It happened about eight months after the Larry Parnes audition in Liverpool, and in this time the band had played shows all across the United Kingdom and in Hamburg, Germany. When they left for Germany, they were still pretty much amateurs, but the band that returned to England in the winter of 1960 was tight, ambitious, and professional. They came back with a much larger repertoire and greatly improved musicianship. A fellow musician described the transformation: "They wore black leather, had brand new instruments and played brilliantly."[16] They also played loudly and with a confidence that had been sorely lacking in the Quarry Men and Silver Beetles, a confidence that impressed fans, journalists, and fellow musicians alike.[17]

Both John Lennon and George Harrison felt that this was the high point in the development of the Beatles as a rock band. When the group left for Hamburg, the Shadows were top of the pile of English pop music, and most of the Liverpool beat bands copied their sound as well as their stage outfits and guitars. The Shadows had helped inspire the crowded beat scene, and it was hard to get away from their sound of three electric guitars and drums. The Litherland Town Hall billed the Beatles as straight "from Germany," and the fact that many in the audience thought them to be foreigners shows how much their music differed from the other Mersey Beat groups.

It was not a good night for a concert. Tony Bramwell remembered it as freezing and snowy, and contrary to the legend, not that many people were there.[18] As soon as the band started up with a blast of amplified guitars, the crowd rushed toward the stage, and the excitement never let up. Brian Kelly, the promoter, was impressed; this was the loudest group he had heard, and he recognized that their "pounding, pulsating beat . . . would be big box office." One fan remembered them as young, rough, and sexy.[19] Even though their leather outfits were now considered low class, they added weight to the contrast with the Shadows, whose dinner jackets now looked distinctly unexciting. The Beatles came across as wild and unrestrained, especially when they incorporated the little vocal tricks and slurs they had picked up from their store of R&B records. This was the big difference, and as George Harrison concluded, "that's why we became popular."

The Beatles were able to stand out among the numerous guitar bands in the city because they kept true to their American inspiration: in 1960 the Beatles were rock purists. In John Lennon's opinion, when it came down to playing "straight rock," nobody in the country could touch them. In a column in *Mersey Beat*, Bob Wooler said that many people had approached him after the Litherland concert and asked him why the Beatles had become so popular. His answer was that the band had resurrected the excitement and rebellion of the early years of rock'n'roll and had "exploded on a jaded scene" dominated by pale imitations of Presley and play-by-the-numbers Shadows clones.[20] The Beatles were, in the words of Ian Samwell, closer to the real thing than anyone else around. Their copies were more attentive and respectful to the American originals. When the Rolling Stones challenged their popularity a few years later, they did it on the same grounds; they were closer to the American masters than the most successful band in the world, which had sold out and gone pop.

The crowd at Litherland Town Hall was not an easy audience. Like the venues in Bootle and Garston, it was notorious for fights, and even the young ladies in the crowd put fear into the Beatles' entourage. Yet Litherland was a triumph and marked the beginning of a new phase for the band. John remembered, "That's when we first stood there being cheered for the first time."[21] Brian Kelly immediately booked them for scores of gigs over the next few months, and they hired Neil Aspinall to be their full-time road manager. The crowds were getting bigger and more excitable. At a return engagement at Litherland Town Hall on Valentine's Day, 1961, the bouncers had to protect the band as the audience rushed the stage. Many more girls were hanging around them, and soon they moved from steady girlfriends to playing the ever-growing field.

During 1961 they competed for birds and audiences in the frantic Liverpool beat scene, which consisted of at least three hundred amateur and semiprofessional groups. The enthusiasm for rock'n'roll was strong in Liverpool, and there was a great deal of scouser pride in the size and vitality of the local music scene, which encompassed not only beat groups but also folk, blues, jazz, and R&B. The new music coming from America had swept up the large Afro-Caribbean community as well, and there were probably more interracial bands—such as Derry and the Seniors—around Merseyside than in any other city in the country. Rock'n'roll might have made the most noise, but there was still enough support to maintain a thriving jazz scene on Merseyside, which encompassed everything from trad jazz, New Orleans style, to cool jazz, the latest musical import from America. By the end of 1961 the Beatles were acclaimed as the best band in Liverpool in a vote organized by *Mersey Beat*. They were on their way.

BEING DIFFERENT

Just like the crowd at Litherland Town Hall, the American fans found the Beatles' music loud, fresh, and different—a remarkable accomplishment if you consider that much of their repertoire at the time consisted of covers of American records. But what struck Americans when they first heard the Beatles was the newness of the sound. The consensus among fans everywhere was that the Beatles sounded different; "fresh" is the adjective that crops up again and again. They had the same effect on the engineers when they first played at Abbey Road: "It was the freshest music I'd ever heard," one said. The historian of rock'n'roll Charlie Gillett thought that the vocals made the Beatles sound different to English listeners. The vocals were a new combination of two American styles, the hard-rock style of singers like Little Richard and the call-and-response, gospel-tinged style of the girl groups.[22]

Since many Americans knew nothing of the rockabilly and R&B roots of rock'n'roll, the Beatles' early repertoire must have sounded refreshingly different to their ears. Its delivery, in scouse-inflected English, stood somewhere between the perfect enunciation of Julie Andrews and the American ersatz of Cliff Richard—the opposing poles of English pop music as discerned by American listeners. The Beatles might have incorporated Americanized pronunciation and rock or soul vocalizations for effect, but they kept to their own voices without succumbing to the temptation to impersonate Americans in the grand old English music hall tradition. This made an important difference, as a fan explained: "Everything was so new, between the British accents and a different beat, and different lyrics, especially."[23]

Nevertheless, parts of the American press corps, including *Newsweek,* found the band's sound "achingly familiar," and both American and English record companies came to the same conclusion: the guitar sound was out of style, and the Beatles did not bring anything new to pop music. Certainly John Lennon's harmonica work added a lot to their early hits, but was it markedly superior to Delbert McClinton's piece on Bruce Channel's "Hey Baby" or that of any of Bob Dylan's early records?[24] Their harmonies were beautiful, but were they markedly better than those of the Everly Brothers—whose records the Beatles copied and whose sound many people heard in the first Beatles records? The Beatles did play with energy and excitement, but they were far from the only band resurrecting the good old days of rock'n'roll. Was their "Twist and Shout" rawer or more aggressive than the Kingsmen's "Louie Louie," a record that exploded onto the rock scene in 1963, or louder than the Kinks' "You Really Got Me," which amazed English record buyers in 1964? The Beatles were not the only guitar band that played R&B covers in an African American style, but somehow they made it sound original. At the same time black musicians credited the Beatles with introducing their blues and R&B to the white audience in America, the kids in the crowd were being swept away with its novelty: "It was a totally new style of music, totally exciting," said one American fan. Another said: "They're different! They're so different!"[25]

The look played a large part in this perception of difference. In England the Beatles differed from the nondescript Shadows, and in the United States they contrasted with perfectly groomed pretty boys crooning romantic ballads on *American Bandstand*. Rock music was as much seen as heard, and this was especially true for the Beatles. The English journalist Maureen Cleave concluded that it was "the looks that got people going."[26] The 1963 *Newsweek* article that introduced the band to American readers pointed out that the music was "even more effective to watch than to hear. They prance, skip and turn in circles."[27] If a fan accessed new music by listening to it across the vast American airwaves, he or she might not have received the Beatles' sound as original and fresh, but on television the band came across as different. This sense of newness was linked to the perceived foreignness of the Beatles and the innovative ways in which their carefully crafted image was presented to the fans.

CHAPTER 7

THE LOOK

Brian Epstein became the Beatles' manager with no prior experience in artist management. He was an amateur and ready to admit it. His musical tastes were far removed from those of the Beatles, and he originally pushed them to perform and record middle-of-the-road pop songs and dance-band standards. Although an expert on record retailing, he was a stranger to the recording process and had no knowledge of the technology. He had no professional contacts with impresarios like Bernard Delfont or Lew Grade, who dominated live entertainment in England from their bases in London.

Yet Epstein's youth, Jewishness, and homosexuality opened doors for him in the world of rock'n'roll management that was young, Jewish, and often homosexual. Joe Meek and Larry Parnes were gay men, as was Norman Newall, a leading record producer for EMI, who was called the "queen of Denmark Street," where record companies, studios, music publishers, and musical instrument stores came together. As Paul McCartney said later, there was "a gay network in show business" that encompassed venue booking, artist management, recording, song publication, radio, and television production.[1] All were successful because they had the knack of indentifying pretty young men who would appeal to the young girls—the critical audience for pop records.

What Brian Epstein brought to the table was his knowledge of the record industry, his knack for picking out a hit record (something he shared with George Martin), and an expertise in what he called "presentation." He was a success at running a record store because he took the time to understand what his customers wanted. He also brought his fashion sense and an understanding of how style interacted with show business. In his account of the first meeting with the Beatles, Epstein made no comment about their music but talked about their dress and their group personality, and these were the elements he would shape. The story goes that he famously cleaned them up and thus prepared them for stardom, but the process of creating their look was more complex than just putting them into suits and trimming their hair. Epstein opened many doors for them, bringing his knowledge of the outside world to provincial Liverpool. He saw the Beatles as a "theatrical attraction" rather than a guitar band, and thus fashioning a look was important. Epstein's own impeccable English taste, his theatrical background, and his refusal to behave like the other managers were important assets for the Beatles.

Everyone who met Epstein commented on his immaculate clothes and grooming, which signaled taste and breeding. He might have worn a bit too much aftershave, but his suits were expensive and well tailored, and his accent (called "Liverpool posh" by scousers) rarely failed to impress. Brian did things with a style that won over the individual Beatles and their families. An example of this is his choice of transportation when they went to Hamburg right after signing a management contract with him. The first time, they had gone on a crowded boat ferry in an old van. But now they flew to Hamburg, a big deal at a time when air travel was considered a luxury. Ringo was at that point still playing in Rory Storm's band and they had already taken the first-class route to Hamburg: "No van for us—we had the suits—we went by plane, which was a thrill."[2] The Beatles, their families, and their fans were equally thrilled by the new professionalism of the band.

The suits favored by Rory Storm and the Hurricanes came in bright neon colors and were complimented by string ties and white buck shoes. Comparing these suits to those approved by Epstein for the Beatles tells us a lot about his sense of style. The Beatles were not averse to matching outfits on stage, for this was a sign of professionalism (and if Ringo is to be believed, it justified greater levels of luxury on tour), but they were attached to the flashy stage outfits associated with rockabilly and R&B acts. They loved the leathers, but as soon as they brought them back from Hamburg, they received unfavorable feedback from audiences and promoters. Even in the 1960s dress codes were rigorously enforced in English entertainment venues, with jeans and leather jackets strictly prohibited. Rock'n'roll outfits smacked of juvenile delinquency and signified lower class—the two great fears of venue promoters—and the Beatles admitted that they looked like hooligans.

The suits that Epstein chose for them were strongly influenced by the latest European fashions, the styles he had picked up on his visits to London or the continent. They were not the bland, roomy American tuxedos favored by the dance bands, nor were they the outrageous stage clothes modeled by Rory Storm and the Hurricanes. The luxuriant material in conservative browns or grays came from Italy, one of the favorite inspirations of the young men in London who were developing "mod" fashions. But the almost luminous mohair material reminded the Beatles of the sharp R&B groups they admired. (Had they been jazz fans, they would have also noticed that many of the young, hip players also favored this style.) The cut of the suits also came from Italy—the slim look with narrow collars, often highlighted with velvet trim, and the elongated, uncuffed trousers that stressed the vertical lines of these rail-thin young men. The trousers were more elegant than the tight drainpipe styles the Beatles struggled to get into while playing the joints of Hamburg. The clothes Brian Epstein bought for them created a modern look that was far from the norm of American pop entertainers, who preferred boxy suits with wider, more angular lapels. The effectiveness of Epstein's choices can be seen in the reaction to their Ed Sullivan broadcasts, when many viewers were taken by the Beatles' stage outfits: "They were so unusual looking because they had these little suits on and they had their long hair." Along with the Beatles' British accents, the suits and hair gave an impression of freshness and novelty, contributing to the sense of newness that was the core of their appeal at the beginning of Beatlemania.[3]

The famous collarless jackets were inspired by the designs of Paris couturier Pierre Cardin. John and Paul first saw them when they visited Paris during a short holiday in 1963. They were now aspiring to what they called an "arty" look, which owed a lot to European style, but the tailors who actually made these famous suits, D. A. Millings and Son of Old Compton Street in Soho (only a few doors down from the 2i's coffee bar, which claimed to be the birthplace of British rock'n'roll), already had experience in making similar jackets for stewards (and Cunard Yanks) on ocean liners. Dougie Millings and the Beatles collaborated in the design of the distinctive jacket. After the Beatles wore them on their television appearances, a stampede for collarless jackets started on both sides of the Atlantic.

The Beatles made it clear to both Millings and Epstein that they did *not* want the same suits as those made for Cliff and the Shadows. They always wanted to be different, to stand out in the herd, and after the runaway success of what were soon called Beatles jackets, the band worked with the tailors to come up with something completely different for their next outfits. The Beatles made zippered Italian Chelsea boots fashionable in England and did the same for buttoned-down shirts and narrow knit ties. Their look had an immediate effect on the young people with whom they interacted. An engineer at Abbey Road recalled, "What struck me most about the Beatles when I first saw them was their skinny knit ties . . . within a short time, it seemed like everyone at EMI was wearing them."[4]

The final, and most important, element in their look was the haircut, the famous Beatles mop top that is so well known it needs no description here. Although the bands that followed them had much longer hair, in the early 1960s the impact of the Beatles' hair was much greater. The hairstyle was inspired by Astrid Kirchherr, the German girlfriend of Stu Sutcliffe and an excellent photographer, who created some defining images of the Beatles. The haircut reflected French and German styles associated with college students and intellectuals, the exis. Paul made this very significant statement in the Beatles' argument with George Martin about recording "How Do You Do It": "We're students and artsy guys—we can't take *that* song home to Liverpool, we'll get laughed at." In Hamburg, exi hair was considered a statement, as Jurgen Vollmer pointed out: "To comb your hair forward instead of back like everyone else did was to express an alternative lifestyle and the first hairy stroke against bourgeois society."[5]

The European influence on the Beatles made them fashion leaders rather than followers. At home they were considered mods, a look inspired by Italian fashions and continental ideas of cool. In America they looked European and different. It might have been said with some hindsight, but record executive Alan Livingston summed up the marketability of the band perfectly: "Here was a different sound and of course the boys had a different look and had great promotional possibilities in the teenage market."[6] The impact of the look was reinforced by Epstein's insistence on uniformity. They always dressed alike, with matching suits, shoes, and haircuts, and he was known to throw a fit if Ringo turned up without a tie and spoiled the look. Brian considered every aspect of their presentation; not only did he want the Beatles in identical suits, he also wanted the same uniformity in the equipment they used on stage and thought about painting all their guitars black (presumably to match Ringo's drum set). He even asked the boys to cut off the ends of their guitar strings to make them look neater!

The look was what Brian was all about. When Billy J. Kramer wanted approval to buy a shirt in a New York store, Epstein refused: "Because it's not your *image*, Billy."[7] One of his first acts as manager of the Beatles was to approach professional photographers and build up a portfolio of images. Epstein ought to be remembered as a pioneer in photographing popular musicians, for under his tutelage the Beatles moved out of photographers' studios to be presented against grungy, industrial landscapes.[8] From that point on it became almost obligatory to photograph rock bands amid rubble. Brian Epstein didn't just clean up the Beatles; he softened their image to appeal to more girls, especially the younger ones, who might have found Elvis with his Latin looks and white-trash background a little scary. The Beatles had little of that danger, and the friendly, accessible mop-top look attracted teenage girls. The androgynous hair was not as provocative as the slicked-back "rocker" hairstyles inherited from Elvis and rockabilly guitarists. The Beatles were irreverent but never dangerous. They managed to get away with the duality of looking and playing rough but coming across as safe. The girls loved them because they were cute and friendly, but the boys saw them as tough. They hinted at rebellion but played it for its humor rather than as a biting critique of the establishment.

The key to understanding the Beatles' style was that they always wanted to be different, and this covered the music, attitude, and appearance of the band. The look was even a consideration in choosing guitars as they were often captivated by the shape of the instrument and the effect it would have on other players or the audience. John Lennon's famous Rickenbacker 325 is probably the most valuable guitar in the world. John bought it while the Beatles were playing in Hamburg, after he saw one on an album of the George Shearing group, with Jean "Toots" Thielemans on guitar. The 325 was strange and exotic-looking, with a futuristic shape and exaggerated "cresting wave" cutaways. John told Toots Thielemans that any guitar good enough for George Shearing was good enough for him, but this was just scouser bravado. Pete Best probably came closer to explaining John's motivation when he remembered the impression it made: "People were like: My God, what's John playing? We've never seen anything like that before."[9]

Appearance also played a part in Paul's choice of guitar. As Paul assumed bass duties in Hamburg, he saw a Hofner 500 "violin bass" in a shop window. It was much smaller than the average bass guitar, and its hollow body followed the shape of a violin. In addition to its striking looks, the 500 was light and produced a woody, rich tone—perfect for Paul's animated performance style. He ordered a custom-made model from Hofner, strung for his left hand, and the instrument was made at the company's factory in Nuremberg and shipped a few hundred miles north to Hamburg. The small body makes the neck look longer, and Paul played it high up, emphasizing the vertical. These two distinctive guitars have become part of the Beatles iconography, and as Paul McCartney pointed out, the Hofner became "kind of a trademark" of the band. When *The Beatles: Rock Band* computer game was introduced in 2009, the marketing put great emphasis on these "trademark instruments," which were nearly as famous as the music itself.[10]

The Silver Beetles made do with old guitars and badly wired, aged amplifiers, sometimes standing on plastic amp covers to insulate themselves from electric shocks. Their equipment did not impress the technicians at Decca or EMI, who used words like duff, ugly, and noisy to describe it. Epstein stepped in and fi-

This shot of a concert at the Majestic Ballroom in Birkenhead, just across the Mersey from Liverpool, in July 1962, shows the Beatles before they became famous. Brian Epstein had become their manager, and now they were playing bigger auditoriums. In the background Pete Best plays drums (Ringo did not join the band until the end of the summer). The photo documents the distinctive shape of their guitars: Paul's Hofner 500, John's Rickenbacker 325, and George's Gretsch Duo Jet. The guitars became a visual shorthand for the band; their images figured prominently in marketing Beatles products. (Courtesy PhotoOffice de Frank Seltier)

nanced new equipment for the band. He made a deal with the Vox company of London to use their amplifiers exclusively in exchange for free equipment and special attention. The Shadows used Vox AC 30s, which the Beatles adopted too—tan-colored amps with the "TV front" popularized by Fender, in which the speaker grille, with its rounded edges, mimics the shape of a television screen. At the end of 1962 Epstein sent the AC 30s to be refurbished and refinished in black. The Beatles were delighted that they had the only black Vox amps in existence, but Vox had started to build them in small numbers, and other Liverpool groups acquired some, to the consternation of Brian and the boys.[11]

Being different was the force that drove the Beatles to collect obscure American records and travel outside the UK. Distinguishing themselves from the crowd grew ever more important as they grew famous and attracted a mass of imitators, and this weighed heavily on the direction they took their music. "They'll never be able to copy that" as they left the studio was the appropriate praise for any recording that went particularly well. In the good old days, Brian Epstein or George Martin restrained them when they threatened to be *too* different. This was especially important in the American market, where there was already a reaction to the sexuality and aggression of rock'n'roll despite every effort of the industry to keep it looking squeaky clean. The length of their hair was immediately seen as a provocation by parents, especially by fathers, but the Beatles managed to win them over. Their management was so successful in containing the musicians' individuality and candor that it was not until 1966 that they crossed the line and gave their critics an opportunity to attack them (after John said that the band was now bigger than Jesus). A fan summed up Epstein's achievement in one sentence: "This is the first time I've gone nuts over a singer that my parents didn't tell me it was disgusting."[12]

BEING THEMSELVES

Brian Epstein understood that the appeal of the Beatles was in four different but harmonious personalities that came together in a sum greater than its parts. He said after that first encounter in the Cavern, "Each had something. They were different characters, but they were so obviously part of the whole . . . There was something enormously attractive about them."[13] Epstein nurtured the northernness of his charges and allowed them to retain their identity as Liverpudlians, which he had discarded years before. Rather than transform them, in the manner of other managers, Epstein let them be themselves while gently constructing an image reflecting the group dynamic. As he always took pains to point out, the

Beatles were different from the rest of England's rock'n'rollers because they were authentic—they were themselves.

Epstein and his staff had a difficult task to convince fans that fame and fortune had not changed the members of the band, because the media constantly circulated stories about their fabulous wealth. The Beatles were famous primarily because they were rich and successful. The official history summed it up: "Four working class Liverpool lads who in four years became millionaires and the best known people in the world."[14] The first news articles focused on their wealth and their new homes and cars, describing in detail each of their Aston Martins and Rolls Royces. George Harrison acknowledged the speed of their ascent when he admitted, "It's like winning the pools"—the weekly bet on soccer games that most British males made in hopes of winning millions of pounds.[15] This was a new sort of fame. Rather than politicians and royalty, the people who had won the pools were now on the front pages of the newspapers, especially the tabloids.

The most-asked question of the Beatles was how much stardom had changed them. This was true of American and European interviewers. The answer was, not at all: Paul had not changed ("he was still the same person"), nor had John, George ("a modest man"), or Ringo, as his mother assured *Mersey Beat*: "They're still the same as ever and quite sensible with all the money." The Beatles told the fans, "None of us look upon ourselves as stars."[16] They appeared to remain the same working-class scousers who had started the band. This image went over the Atlantic with them, and fans in unfashionable American cities saw some resemblances: "We thought of Liverpool as a dirty coal town like Birmingham [Alabama] had been in the past." The fans in Cleveland associated with a grubby town on a river: "Liverpool? It's like Cleveland but much worse." The Beatles seemed to be not that much different from the people in the audience: "I saw them as working class, normal guys like me. They seemed grateful to the fans, appreciated us. They had a little band that got big."[17] As Michael Frontani pointed out, "The Beatles' ordinariness was a cornerstone of their image."[18]

The Beatles started as a local band with strong ties to the community. They were able to generate unusual devotion and loyalty from their Liverpool fans, who thought of them as *their* band. The four came across as sincere, unpretentious local lads, who spoke the language of the fans—not just the broad scouser dialect that John and Paul exaggerated when they appeared on the national media, but the intonation, slang, and slurs of their audience. They were "the utmost, ginchiest skizziest and craziest" of the cool cats in Liverpool, according to a fan.[19] They stopped the show to talk to or to abuse their friends in the audience at the Cavern Club, and as they performed to larger groups, they still managed to maintain

personal contact with the crowds, talking to them after the show, signing autographs, and perhaps dallying with the cute birds. They hardly changed their
demeanor when they played large auditoriums in the United States. Even though
it might have been delusional, the fans felt this closeness—"Oh my God! My girlfriend kept saying: Ringo's looking at me! Ringo's looking at me!"[20] Throughout
their long careers, the Beatles made the effort to stay close to their fans, even if
they feared them occasionally. As John Lennon said, "We believe in our fans . . .
Sometimes we have to turn and run . . . because we don't want to cause riots."[21]
It was Lennon's policy of staying accessible to the fans that eventually cost him
his life.

The Liverpool fans had a proprietary relationship with the group. It isn't
usual for fights to break out in the audience over a personnel change in a rock
band, but so fervent were the supporters of Pete Best that the Beatles and their
manager were under physical threat when they replaced him. For the faithful, the
Cavern Club was the true home of the Beatles. As the band grew more and more
popular in 1961, the fears that they would leave also grew stronger: "If you have
a hit record you'll go to London and we won't see you any more," "you gorra stay
in Liverpool. You're ours!"[22] This fear was real, for as the records achieved chart
success, the move to the entertainment capital became more likely. The letters
the "hurt and disappointed" fans wrote to *Mersey Beat* grew more anguished as
the fame of the band grew: "They are not our Beatles any more," one wrote. Another reassured her colleagues that although the fan club had moved to London,
there was still a Liverpool branch, and the Beatles were not going to leave home
like the other stars.[23]

Once the Beatles moved to London, they continued to pledge their loyalty
to their fans in Liverpool. Their message, "We'll be back," was repeated when
they started to tour the United States, but now "home" was the whole kingdom
rather than the clubs and suburbs of Liverpool that claimed them. The broadcast
interviews that the Beatles sent from their hotel in New York were similar to
those they had sent from Epstein's offices in London, affirming their loyalty to
the fans and the home they had left behind. They made several long-distance
telephone calls to English deejays (either live on air or taped) to confirm that they
were returning home soon. Their management was worried that the fans would
feel neglected in the Beatles' absence and shift their allegiance to other bands.
This reflected the special conditions of the market for pop music, where fame
was short and brutal, but it also might have been a heritage from the old days of
music hall, when an American tour might take performers away from home for
months rather than weeks. The conventional wisdom in the pre-Beatles era of

pop music was that popularity and sales peaked quickly and then vanished. This process framed the approach of record producers, concert promoters, and management. Capitol's president Alan Livingston was stunned by Beatlemania: "I mean it happened so fast. It happened overnight."[24] This fundamental law meant that management had to exploit success quickly with nonstop touring and record releases, while reassuring the audience that the band members "never want to forget what they owe to their fans." The fans listened and took it to heart. On the day the Beatles announced their breakup in 1970, the fans were devastated. When interviewed outside the Apple Corps office in London, one said, "We grew up with them—they belong to us."[25]

MANIPULATING THE MEDIA

The construction of a marketable image for the Beatles required the cooperation of the mass media. The Beatles were media friendly, and their skill in dealing with them was an important element in their success. They managed their relationship with an English tabloid press that was prepared to savage them once they smelled blood. The ease with which they circumnavigated the traps and provocations that had been prepared for them amazed the press corps on the band's first American tour. In contrast to stars like Elvis, who had been protected from the press by management, the Beatles sought out journalists (with a little prodding from Brian Epstein and Tony Barrow) and engaged them with honesty and charm. It was understood among the journalists who met them at Kennedy Airport that this was going to be a hatchet job, they were going to "kill" them, but the Beatles turned this around and made the press their allies in their invasion of the United States. John Lennon thought that mastering the tough English press prepared them for the American tour: "We learned the whole game. When we arrived here we knew how to handle press." It took many years after their American tours before reports of their ungentlemanly behavior began to seep out. Their management thought it was a miracle that none of the orgies, groupies, and paternity suits got into the press. Once Beatlemania had taken hold, the press had an investment in the band that they were loath to damage. As Peter Brown pointed out, they had a "healthy dependence" on the Beatles.[26]

Radio exposure played a big part in Epstein's plan to turn his uncouth quartet into the greatest theatrical attraction in the world. As George said, "The radio was the thing in those days." It had the ear of the nation and was by far the best way to market new music. During the first few months of Epstein's management, he sent applications to regional radio shows in the Northwest for auditions. The

Beatles made their first radio appearance in March 1962, on the BBC's *Teenagers Turn (Here We Go)*, a local talent show—programming that went all the way back to the beginning of commercial radio in the 1920s. Their big break was appearing on *Saturday Club*, the BBC's main outlet for youth music, which was incorporated into the lifestyle of nearly every teenager in the United Kingdom. It went on the air at 10 a.m. Saturday, early enough for some (including Paul McCartney) to listen to it in bed. The Beatles became accomplished broadcasters, and during 1963, they made around forty appearances on national radio, including a thirteen-week run of their own radio show, *Pop Go the Beatles*.

The Beatles' relaxed style, quick wit, and easy rapport with one another went over very well on the radio. The banter with friends in the front row of the Cavern was drowned out by the screaming when they played larger auditoriums, but radio gave them the chance to reestablish this sense of friendly familiarity with their fans. Radio created an aural picture of four separate identities within the group, with different voices and laughs, different levels of reticence or boldness, and slowly evolving roles: John the leader, Paul the heartthrob, George the quiet one, and Ringo the clown. Their interviews in teen magazines and radio broadcasts provided enough information to give the fans a complete picture of each member of the band. Jonathan Gould imagines America at the height of Beatlemania with a million girls picking their favorite Beatle after months of consuming tidbits about them in broadcasts and press articles, then conferring with their friends and parsing the lyrics of the songs for clues. Choosing a favorite Beatle, Gould argues persuasively, defined the fan and her relationship with her peers.[27]

If you listen to recordings of the Beatles talking on the radio, you might get an idea of what Brian Epstein saw in them in the Cavern in November 1961. There was no trace of the artifice that surrounded the staged creations of the rock Svengalis. There were no strained silences. Instead the Beatles played themselves. Three out of four of them kept their own names, and all of them managed to come across as real. In acknowledging the role that radio played in the presentation of the Beatles "being themselves," Derek Taylor included the holiday messages and interviews it broadcast as part of "an honest paying of all the expected dues."[28]

TELEVISION

Radio could conjure up an aural portrait of the band, but television presented the all-important look, the complete image constructed by Brian Epstein. The

Beatles grew up listening to the radio, but most of their audience was weaned on television. While Americans were enjoying variety shows on their cheap Sears or Zenith television sets, Britons were still awaiting the end of rationing of essential foods. But in the late 1950s the British television audience began to grow. In 1957 only 20 million households in the United Kingdom had TVs, but by 1960 three-quarters of them had one. The most popular light entertainment programs, like *Sunday Night at the London Palladium*, put out by Britain's commercial network, ITV, would be watched by half the population. During the last years of the 1950s, the end of rationing combined with rising incomes and low unemployment led to a surge of consumer spending, with expenditures on refrigerators, vacuum cleaners, and televisions increasing 70 percent from the midpoint of the decade—a process supported by the introduction of credit and installment payments.

This was reflected in the growth of the Epstein family business in Liverpool. When Brian's father, Harry, had taken it over from his father, he established another store in downtown Liverpool that dealt in electrical appliances like washing machines, televisions, and radios. The record retailing part of the business took off when Brian was given control of it. At the old Walton store, weekly sales of records hardly reached seventy pounds, but in the downtown store on Charlotte Street, sales were more than twenty pounds every day.[29]

Much of the early television programs came from radio, including the variety and minstrel shows taken from the entertainments of the previous century.[30] The first commercial television broadcast in England came from the Alexandra Palace in 1936 and showed variety acts, such as jugglers, dancers, and comedians. Television kept the music hall alive in England while it was stealing its audience. Vaudeville survived in American TV in programs like *The Texaco Star Theatre* (with Milton Berle) and *Toast of the Town,* introduced by Ed Sullivan. In the 1940s and 1950s, the latter became the highest rated variety show on the air as *The Ed Sullivan Show*. In his long career on television (1948–1971), Sullivan became a media personality and one of the most powerful star makers in popular music.

In the UK, television, like all mass media, was based in the capital, and it was particularly difficult for Liverpool acts to be accepted there. When Sam Leach, a Liverpool promoter and erstwhile manager of the Beatles, tried his luck with the influential London agent Tito Burns, he was told, "They've got no chance coming up here from the sticks."[31] All the television studios in the country were in London, as were all the recording studios. The establishment of the Independent Television authority (ITV) in 1955 provided a competitor for the BBC, and the Beatles owed much of their media presence to the institutional rearrangement of British television in the 1950s, which helped move production facilities out of London to the provinces. The group that won the ITV franchise for the Northwest, which contained Liverpool, took the unprecedented step of building a television studio in Manchester (although it later found it had to have production facilities in London or miss out on a great deal of talent).

Granada television emerged from the Granada chain of film theaters. Its application for a television license played heavily on a focus on the Northwest, its people, and its culture. Granada TVs Manchester studio was the springboard for the Beatles to engage the television audience, beginning with a program called *People and Places* in early 1963, and from then on Granada and ITV played an important part in presenting the Beatles to the nation. The Beatles traveled the short distance to Manchester quite a lot in 1962 and 1963 to work in radio and television studios there—the shift of activities from Liverpool to Manchester preceded the band's important move to London.

The BBC engaged popular music in the manner of an understanding adult who barely tolerated the loudness and impetuousness of youth. Television programming for teens at this time was "something with mountain climbing for boys, fashion for girls . . . that sort of thing," presented formally from the studio in the same dignified manner that characterized British television news.[32] Commercial television was focused on entertainment rather than education, and this made it more tolerant of rock'n'roll. Associated-Rediffusion broadcast *Cool for Cats* from 1956 until 1959, when the Quarry Men were formed and the main vehicle for rock music was American films. Among the many people who felt the impact of *Blackboard Jungle* in English theaters was Jack Good, a producer for the BBC's Light Entertainment television. Impressed by the excitement of teenagers who danced in the aisles to "Rock Around the Clock," he began to think about ways to transfer this energy to the small screen: "I saw the light," he said later. Good convinced his bosses at the BBC to try a new kind of approach that would break all the rules for youth television: "I just wanted wham bam de boo bop, do wop bam boom . . . Tutti Frutti . . . then the next number, next number . . . and bored with that . . . next number."[33]

The *Six-Five Special* debuted five minutes after a brief six-o-clock news broadcast on Saturday, February 16, 1957. It was an experiment that lasted for only six weekly episodes, but it spawned a host of imitators that would change the face of teen programming and become a force in the marketing of pop music. It showcased a broad selection of popular music along with sports and general interest features, at first featuring skiffle and trad jazz, and then moving with the music, from Lonnie Donegan to Tommy Steele. *Six-Five Special* was broadcast live from a studio that had been stripped of sets to hold a melee of teenagers dancing to live music in front of a live audience. Good swung the camera around groups of kids dancing energetically and let their excitement power the images. The frantic pace of his shows, which hardly gave the host and the audience a chance to take a break between acts, presented a sharp contrast to the leisurely procession of variety acts that characterized English light entertainment of the 1950s. The BBC was uncomfortable with the *Six-Five Special* and preferred to stay with the broad magazine format, but Good wanted to concentrate on the music. A new ITV licensee, ABC/ATV, immediately signed Good, introducing *Oh Boy!* in 1958 with instant success. The *Six-Five Special* and *Oh Boy!* produced exciting live television that immediately connected with the large youth audience.

These shows gave performers national exposure and made Jack Good one of the most powerful men in English popular music. His role in making "Move It" a hit was no coincidence. Singers and groups were now sending their demos to Good rather than to Larry Parnes or Bernard Delfont. Michael Cox, a student at Quarry Bank at the same time John Lennon attended the school, got a spot on the *Six-Five Special* in 1959, and this led to a record deal with Joe Meek's Triumph label and the hit "Angela Jones" in 1960, after he performed the song on television. The next generation of televised pop music reflected radio's preoccupation with the charts and its dominant Top 40 format, showing how closely the media was linked with the record business. In the United Kingdom programs like *Pick of the Pops* and *Juke Box Jury* brought everybody into the game by making participants and audience evaluate new records. *Thank Your Lucky Stars* (1961) was commercial television's response to the success of *Juke Box Jury* on the BBC, and by the time that the Beatles were ready to break out of the Liverpool beat scene, it was the premier televised showcase for British youth music.

Once the *Six-Five Special* established itself on the tiny screen, appearing on television became an accepted part of pop stardom, the holy trinity of records, radio, and television enumerated by George Harrison when he explained the route taken by successful rock bands. In the pre-Beatles era of popular music, only records and radio would have been mentioned. Television was the preeminent vehicle to introduce the Beatles to their audience, and this made their appearance vital to their success.

The power of television to create a hit was evident in the first few months of 1963, when the Beatles were pushing "Please Please Me," the single Parlophone released on January 11. Their first ever television performance was on Scottish television's *Round-Up*, three days before the release date. This was followed by *Thank Your Lucky Stars* on January 13, when they again mimed "Please Please

Me" for the television audience. The Beatles had taped this mimed performance at a studio in Birmingham on January 13. On January 16 in Manchester, they recorded the song live for BBC radio's *Here We Go*. "Please Please Me" entered the charts on January 17, and two days later, when the song was broadcast on *Thank Your Lucky Stars*, sales surged accordingly.[34]

Their television appearances went hand in hand with radio broadcasts. On January 21 the band went to EMI headquarters in London to tape a Radio Luxembourg broadcast, which included "Please Please Me," and the next day they recorded it two more times in BBC studios to appear on the radio programs *Saturday Club* and *The Talent Spot*. They also broadcast an interview on the radio program *Pop Inn* to promote the single. By the beginning of February, "Please Please Me" was in the Top 20, and a week later it was at Number 3. On February 17 they went to ABC's studios in London to record another performance of "Please Please Me" for *Thank Your Lucky Stars*. Two days later Bob Wooler would tell the disgruntled fans at the Cavern Club that "Please Please Me" was now Number 1. On February 20 they sang the song live on BBC radio's *Parade of the Pops*. The promotional effort began to wind down in March, because their next single, "From Me to You" (written on the February 28), was recorded on March 5, but "Please Please Me" continued to be broadcast on ABC's televised talk show *ABC at Large* and on radio's *Here We Go*.

This intensive media promotion brought the song to a broad spectrum of likely buyers, including young children who watched a hand puppet called Lenny the Lion work through the song on the BBC's *Pops and Lenny Show*.[35] As the sales of records increased, so did the media exposure, which in turn drove sales. Television brought the Beatles into the homes of millions of people, and soon they moved from playing clubs in Liverpool to filling film theaters and dance halls across the kingdom. They worked the national cinema and dance hall chains, such as Odeon and Top Rank. The magnificent theaters that had brought celluloid images of Bill Haley and Elvis Presley to British audiences now presented the Beatles in person.

FILM

The importance of the look meant that the Beatles' management considered a full-length feature film as soon as the first records became hits. A film was an accepted part of pop stardom and represented the top of the entertainment business. Rock'n'roll films had changed very little since *Rock Around the* Clock and *Jailhouse Rock*: a thin plot gave the pretext for musical interludes, with awkward musicians pretending to act. The Beatles and their management aimed to make a film that was different, and far more realistic, than the movies of Elvis and Cliff that they had come to despise. "It was all false before," as one of them said later, indicating how much they thought they had changed pop music and its movies.[36] The strategy of being themselves was central to the Beatles' first film, *A Hard Day's Night*. Originally titled *Beatlemania*, it was intended to be a documentary of the band at work. This idea was eventually shelved, but the feature film *A Hard Day's Night* had none of the artifice or commercialism of the rock'n'roll films that preceded it. What made the Beatles' first film different was its authenticity, the impression that this was a faithful and privileged look at the real lives of the band members. *A Hard Day's Night* further articulated the Fab Four's individual characteristics to a much greater audience. As film critic Roger Ebert stated, "After that movie . . . everybody knew the names of all four Beatles . . . everybody."[37]

From the opening shots of the film, when handheld cameras swoop in on London's Paddington Station, there is the impression of a reality that keeps up with the swirling action of Beatlemania. The documentary look of these opening scenes was reinforced by the crowds of young girls who pursued the band and provided the soundtrack of screams. The production crew had planned to use professional, unionized extras, but as soon as the word was out, the film set was besieged by hordes of real fans, and they took over the role of extras— adding a measure of authenticity that could not have been planned. Controlling the fans became a major headache because they disrupted the shooting, attacked the stars, and even used hacksaws and glass cutters to break into the sets. Their intrusions, along with the everyday locations, made the film look real. The camera followed the Beatles as they ran from their fans, and when George accidently fell over, to the roars of laughter from his fellow Beatles, director Richard Lester kept the camera rolling to capture the moment.

A Hard Day's Night was shot in black and white and had a documentary look. Lester's debt to the documentary tradition and to the British Free Cinema movement is evident, as film historian David Cook and other writers have noted. Free Cinema began as a program of short films shown in London's National Film Theatre in 1956, a movement that brought together techniques of documentary realism with a sense of social justice. It was to be personal filmmaking, committed to the dignity of the individual and "the significance of the everyday," and much different from the existing English films that concentrated on London and the affluent South rather than on the dark industrial cities like Liverpool and Nottingham. Tony Richardson and Karel Reisz took their lightweight 16 mm cameras into youth clubs and pubs and made classic documentaries like *Momma*

Don't Allow (1956) and *We Are the Lambeth Boys* (1957). In these documentaries the kids are shown talking, smoking, and flirting, but it is the wild abandon of their dance moves that jumps off the screen, as the music takes them away from black-and-white realities for a moment; this is also the message of *A Hard Day's Night*. When the Beatles, as the oppressed workers, escape from a television studio into the outside world of freedom, silliness, and song, it provides a moment of joyous liberation that critic Andrew Sarris called one of "the most exhilarating expressions of high spirits I have ever seen on film."[38] At the core of both *Momma Don't Allow* and *A Hard Day's Night* is a belief that music can be the means to achieve transcendence.

True to the spirit of Free Cinema, *A Hard Day's Night* shows us everyday life in carefully observed detail, but these are pop stars, not manual workers, and their day's work consists of press interviews, parties, and television broadcasts. This daily grind is a lot different from the dirt and noise of factory work, but it is a grind nevertheless. The film accurately records the work of rehearsing and broadcasting the musical television show—something the Beatles were now experts at doing. They have to deal with various entertainment industry types, from nagging road managers to authoritarian television producers, all demanding and annoying in the ways adults are always depicted in these films. The Beatles could not resist the temptation to bite the hand that fed them: "Susan Campy, our resident teenager," is called "that posh bird who gets everything wrong" by George in a reference to Cathy McGowan, the resident trendsetter on TV's *Ready Steady Go!*—the most popular youth music show at the time. While cleverly satirizing the pretensions of teen television and the fashion industry, *A Hard Day's Night* still maintains the conventions of the teen film—in which carefree young heroes overcome the anxieties and hypocrisy of the older generation. Because Lester had decided to build the film around the real-life experiences of the Beatles, it was appropriate that they should be shown in television studios and the conference rooms of advertising companies.

Liverpudlian writer Alun Owen wrote the screenplay for *A Hard Day's Night*. Owen had made his reputation by portraying northern working-class types in plays like *No Trams to Lime Street* and in television series like *Z Cars*, which were set in the mean streets of Liverpool. His job was to make the dialogue as true to life as possible, and although the four musicians said afterward that they were not given many opportunities to improvise on camera, their lines come across as natural and unforced. The four heroes sound unashamedly northern and ready to thumb their noses at the pretensions of the London establishment and the arty, fashionable set (a set that Paul McCartney and John Lennon were willingly entering on their days off). Their thick scouser accents were stressed on the soundtrack as much as their music, and they treat the capital city as another place to pass through rather than their home. The film underlines not only the conflict between old and young but also southerners versus northerners, with the latter down to earth and far more sympathetic.

As far as the management and fans were concerned, the selling point of the film was that it represented the Beatles as they really were, yet John, Paul, George, and Ringo were playing the Beatles rather than themselves. They had had at least two years' experience by the time they went onto the set of *A Hard Day's Night*, filling out the characters that made up the Fab Four in countless radio and press interviews. Many of the people close to them noted that they were acting the part. The Maysles brothers thought that "they felt they had to act for us," and Richard Lester also believed that the Beatles were playing themselves but added, "they were pretty good at that."[39]

A Hard Day's Night succeeded in appearing real. The Maysles brothers' documentary *What's Happening: The Beatles in New York* was actual cinema verité, filmed on the spot, yet it looked so close to Lester's film that Brian Epstein would not let the Maysles release it until *A Hard Day's Night* had premiered.[40] Although *A Hard Day's Night* maintains some continuity with the documentaries of the 1950s, the Beatles' films of the 1960s reflect a different sensibility, a truly sixties aesthetic. The critic Christopher Booker saw the sixties as a time when "the façade had never been so important, or so unreal," and the Beatles films, and the band's look, fit this assessment perfectly.[41]

What *A Hard Day's Night* captured and preserved forever was the Beatle image that Brian Epstein had wrought: four lovable, irreverent mop tops immaculately dressed in fashionable identical outfits. It presented the band in much finer detail than any television show or photograph, and it rounded out the image of the Fab Four already established in the mass media. But like all motion pictures, *A Hard Day's Night* was a fantasy, a portrait of the band through rose-colored glasses, without the cigarettes, bad language, and casual sex. All four of them were heavy smokers. Producer Walter Shenson noticed the Beatles smoking "like chimneys" on the set and tried to get the cigarettes out of their hands before the cameras rolled or clicked—fourteen-year-old girls should not be exposed to that.[42] The reality of the Beatles' individual characters and their decadent lifestyle was not revealed until Brian Epstein's untimely death, and then the often unpleasant truth came out. John Lennon concluded, "We don't put on a false front or anything. But we just know that leaving the door, we turn into Beatles . . . We're not the Beatles at all. We're just us."[43]

CHAPTER 8

THE FANS

Who were the people buying the Beatles records and screaming at their concerts? If we look at the pictures of the crowds waiting in line to get into the Cavern Club in downtown Liverpool in the early 1960s, we see a group of well-dressed young adults rather than teenagers. These performances were lunchtime sessions put on for the workers who left offices, warehouses, factories, and retail shops to enjoy an hour or so of music. Predominantly female, they had recently left school and were now entering the workforce—just like the four young men who were entertaining them. Their peer group, as described by Cynthia Lennon, were office clerks, shop assistants, factory workers, students, and layabouts, many of whom had just reached the peak of their disposable income before marriage and house purchase eroded their spending power. When English people and newspapers talked about teenagers, they invariably saw them as employed rather than students or layabouts: the *Daily Mail* thought of the "typical" teen as a hairdresser, typist, or secretary.[1]

In London the Beatles attracted a younger audience. The girls who chased them around in *A Hard Day's Night* were school aged, as were the fans in the pictures taken of the "siege" of the London Palladium. When the band returned from the United States in February 1964, the police were disturbed by the number of children in the welcoming crowds, especially the eleven-year-olds, whom the police thought were in need of "care and protection" and adult supervision.[2] The crowds of fans who welcomed the Beatles to New York were predominately female high schoolers. A poll of the American branch of the Beatles' fan club discovered that the average fan was between thirteen and seventeen years old, a white, female, Christian, middle-class, B-student who owned a transistor radio.[3] As the press promoted Beatlemania and the band appeared on more television shows, the age of the fans dropped to encompass preteens as well as teenagers. A significant proportion of the fans were chaperoned by parents or elder siblings to their first Beatles' concert. Kindergarten and preschool students were among

Beatlemania began in Liverpool at the Cavern Club on Matthew Street. Here, Liverpool fans— well-dressed men with shirts and ties; most women in skirts and heels—wait to get into a Beatles lunchtime show in 1961. At one of these sessions, Brian Epstein first met the band. Matthew Street is now the center of Beatles tourism, and the Cavern Club is a well-visited shrine. (Courtesy National Museums Liverpool)

the vast audience that saw the historic Ed Sullivan performances; they were able to walk into the sitting room and watch TV but were probably not old enough to go outside unattended. Yet they were also consumers whose parents kept up the flow of income to the record companies and to the army of merchandisers who exploited the popularity of the band to children. Toy makers were a major bene-factor of Beatlemania, especially those who produced dolls and stuffed animals, but even kit makers like Revell got into the act by producing plastic models of the Fab Four. The ABC television network broadcast a Beatles cartoon that was sponsored by a toy manufacturer who used the show to promote erector sets, car racing games, a toy train layout, and several dolls and puppets. Other products, like candy and soap, were tied into the cartoon series.[4]

The Beatles connected with a larger and more affluent wave of baby boom-ers in the United States, especially girls between eleven and eighteen years old. It was the Americans who created the idea of a "teenager" and built up the in-dustries that provided goods and services to them. Teenagers became a distinct group, not the anonymous mass between youth and adulthood, but a separate category with distinct characteristics and needs. The term came into popular use amid the growing affluence of Americans and their gradual move to the suburbs—the breeding ground of the new TV- and car-based culture. The United States had a leisured class of affluent teenagers that could be found nowhere else in the world, and these were the people who made Beatlemania into a spectacle of sixties' consumerism.

One of the first indicators of the growing financial power of teenagers in the early 1960s was the emergence of magazines aimed directly at young women. This business started in England in 1959 with *Boyfriend,* and in the United States it began even earlier with *Seventeen* in 1944, whose first issue told its subscribers, "It is your magazine, high school girls of America." With articles on fashion, food, and boyfriends, it followed the format of women's magazines but with the content skewed towards the affluent teenage reader. Other teen magazines grew out of Hollywood's publicity machines, who distributed photographs, prefabricated biographical information, and tidbits of gossip about their stars in the 1930s and 1940s. The arrival of rock'n'roll and Elvis rejuvenated this business. Jacques Chambrun was so impressed with the enthusiasm of Elvis's fans that he started a magazine called *All About Elvis* in 1957. This evolved into *16,* and under the direction of Gloria Stavers, this magazine established a format that would endure for the rest of the century.

Stavers had started in the subscription department and spent much of her day reading the hundreds of letters pouring in from young women. Most of the correspondence was fan letters sent to celebrities in care of the magazine. Stavers realized that these were young women writing in, mainly girls from nine to twelve years old, and that they were hungry for information and images about their favorite stars. She moved the magazine toward highly illustrated content, with large "kissable" posters, and "inside information" about celebrities, often built around interviews, such as the regular feature "40 Intimate Questions." Not that many Hollywood stars appealed to nine- to twelve-year-olds, but pop music provided enough attractive young men to fill the magazine's pages, and the advent of Beatlemania made the Fab Four the staple ingredient until the Monkees and Paul Revere and the Raiders came along. The potential of the teenage audience hit advertisers "like a bombshell," and in the early 1960s a stream of new weeklies entered the market: *Teen Screen, Datebook, Tiger Beat, Valentine, Roxy, Marilyn, Jackie, Mirabelle, Romeo, Valentine,* and *Honey.*[5] They focused the energy of fan worship onto a number of photogenic stars and competed with one another to find interesting new feature stories about them, such as "Beatles-Elvis: What Really Happened at this Secret Meeting." Although there was a very high attrition rate, these magazines flourished as a profitable halfway point between comics and stories for children and light reading for women.

No one can deny that the Beatles were attractive young men. In the beginning, when the press was not interested in pop music, a teen magazine like *Boyfriend* could easily be persuaded to do a photo shoot with the band, and these illustrated magazines were first to spread the word about the Beatles. Teen magazines had the same symbiotic relationship with the Beatles as the national press; Beatlemania provided the magazines with unlimited content, and in return the Beatles' management could use them to shape the band's image. When the press moved in on a story like John Lennon's secret marriage, Brian Epstein placed a special article in *Mirabelle* to control the damage. Most of the staff writers on these magazines were not much older than the Beatles, and they usually succumbed to their wit and charm.[6]

As the band's popularity increased, so did the number of publications devoted to youth music, like *Rave* and *Pop Weekly*, and to the Beatles specifically, such as *The Beatles Book*, which began in 1963 and lasted until 1969 (as *Beatles Monthly*). This was the first mass-marketed fan magazine, and it reached 350,000 subscribers. Beatlemania promoted the publishing business, but these highly illustrated magazines also provided much of the raw material for fan mania. The fans kept scrapbooks of pictures of their favorite band, along with newspaper cuttings and other mementos, such as concert tickets and flyers—the printed paper that played such an important part in maintaining the presence of pop stars in their lives. Beatles fans built shrines of memorabilia, pictures, and records in their bedrooms and covered the walls with images of the band. Teen magazines were the major source of images for these recontextualizations of the Beatles.

Like many other products related to the Beatles, these magazines were soon traded across the Atlantic, as desperate fans acquired every detail of the band they could lay their hands on. Girls knew the Beatles' birthdays, the color of their eyes, their favorite foods, the singers they admired, and what they looked for in a girlfriend. It was a point of honor among them to collect this vital information. They were aided in this obsession by a transatlantic network of communication and commerce. As Beatlemania took hold, it generated more teen magazines and related picture publications, which grew into a large publishing business. In the early days of rock'n'roll, Colonel Tom Parker had to hawk photographs of Elvis to the fans outside venues—Brian Epstein did not have to bother because he had an international media industry to do it for him.

Although Beatlemania is always associated with young girls, there were many

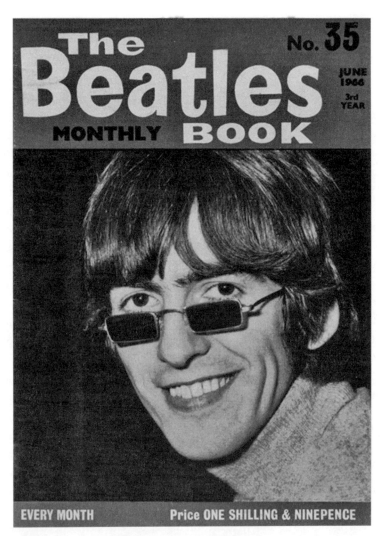

The Beatles
No. 35
JUNE 1966
3rd YEAR
MONTHLY BOOK

EVERY MONTH   Price ONE SHILLING & NINEPENCE

Beatles Monthly (originally *The Beatles Book*), the most important and widely read Beatles fanzine, competed with scores of similar publications but was the only one that could claim to be officially connected to the band through its fan club.

male fans. The pictures of the Cavern Club and other early venues for the Beatles reveal a fair number of boys present. In the crowd taking part in the siege of the London Palladium, the *Daily Mirror* noticed many boys, dressed in "high-necked, tight-fitting Beatle clothes."[7] While the audiences during the first American tour were virtually all female, later tours in 1965 and 1966 brought more males into the audiences. In continental Europe and Asia, audiences were more balanced between the sexes. When the Beatles played in Paris just before they traveled to America, they were appalled to see very few young ladies in the audience! "Where are all the birds?" In Germany they were as popular with men as with women, and one report had boys rather than screaming girls clamoring for them at concerts.[8] The males in the audience were not coming to scream but to listen to the music (or to check out all the girls there). In some masculine circles, it was not cool to like the Beatles when there were more aggressive and louder bands, like the Who, out there, but the golden rule for picking up girls in the early sixties was that you had to be conversant about the Beatles. The music that came across to females as romantic and tender could be interpreted as rough and uncompromising to young males: "There was a camaraderie with the guys there. 'Did you like it?' 'Oh they were tough!' "[9] The Beatles managed to be all things to all people.

People who watched the Beatles perform usually noticed (if they were close enough) that the band always seemed to be enjoying themselves as much as the screamers in the audience, and you only have to look at the marvelous color film of their Shea Stadium concert to witness their joy in performing. Their gales of laughter at someone's mistake always sounded spontaneous and natural—they *were* having fun, just as George had said: "We had fun, you know, we really had fun."[10] Alun Owen pointed to the Beatles' unique ability to convey their great joy in life and in being Beatles to their audience. Their banter on stage, unrehearsed and refreshingly different from the stale platitudes that were the norm for pop concerts, provided a testament to the fun that also seemed embedded in the music. Soon it was part of the act.

BUILDING THE ORGANIZATION

As with everything else in Beatlemania, the establishment of fan clubs occurred at an unprecedented scale and speed. At first the fans in Liverpool either wrote to the band members at home (their addresses were public knowledge) or turned up at their houses hoping for a glimpse or maybe even an audience and a cup of tea. The official fan club started at the Cavern, where the hard core of support first coalesced. Roberta "Robbie" Brown sent out details of upcoming gigs to the

Cavern faithful, and this grew into a club. At the end of 1962 the club reported that a thousand fans had joined since April.[11] Roberta organized a special fan club night at the Cavern that April, called "an evening with John, George, Paul and Pete." In deference to their long-term fans, the band appeared first in their leathers to cheers, then changed back into the suits that had become their new stage clothes.[12] The club began chartering buses to take fans to performances around the Northwest. Soon there were branches outside Liverpool.

The club had started as a purely local effort, part of the personal connection to the band. At a time when British young people did not carry around driver's licenses or credit cards, the fan club membership card was carried in wallets and handbags together with other printed badges of allegiance, such as the Cavern Club card. As the Beatles' fame and audience increased, they appeared less in person and more in printed documents. The fan club mass-produced personal messages and autographed portraits and sent them out in bulk mail deliveries.

Unlike the record, radio broadcast, and teen magazine, the fan club was not a product that commodified the Beatles; it was a service that drew people with similar interests together and helped create a sense of community. But over time it joined the other factories of Beatlemania, churning out the images and data that kept the movement fully fueled, and it was inevitably taken over by the Beatles' management. Brian Epstein was impressed with the loyalty of the fan club members and took over the operation and finance of the club, appointing a member of his staff, Freda Kelly, to assist Robbie Brown. By this time the club was the largest in the country and dealt with the needs of tens of thousands of fans. The letters were arriving at the rate of many hundred every day, and at the height of Beatlemania, as many as three thousand letters came in daily. There were so many letters that NEMS brought in the Beatles' family members and other NEMS employees to work through the sacks of mail. They wrote back, using up packs of five hundred photographs and compliment slips. In June 1963 NEMS enterprises moved to London and undertook a major reorganization of the fan club. It was divided into northern and southern divisions—the northern based in Liverpool and run by Freda Kelly, and the southern based in London under Bettina Rose; beneath them was a network of local secretaries. Tony Barrow took over management of the national organization, and he invented a club secretary (whom he named Anne Collingham), who answered the fans' correspondence.

The fan club now faced the task of meeting the needs of a client base that would have overwhelmed the resources of all but the largest concerns. For example, when George Harrison turned twenty-one, he received thirty thousand birthday cards! The fan club more or less doubled in 1964, and its workload was so large that it took up all the Monmouth Street offices of NEMS, so the press and publicity departments were moved to other office space on Argyle Street.[13] By 1965 it counted about seventy thousand members. Beatlemania greatly increased the burden on the club because soon after *The Ed Sullivan Show*, sacks of airmail started piling up at Monmouth Street. An official fan club in the United States had about sixty thousand members in 1964 and was eventually taken over by the Apple Corps. There were also many unaffiliated clubs, often run by local radio stations. The WABC fan club in New York was receiving two to three thousand letters a day in January 1964. Soon there were clubs in thirty-five countries.

The fan clubs got special treatment from the Beatles. In December 1963 two club conventions, one in Liverpool and the other in London, were treated to their own special concert. Thousands of fans enjoyed the performance, and they also made up the delirious audience for television cameras to record two shows that were broadcast later—in other words, the fans could also be employed in manufacturing the product. By this time the finances of the club were stretched to meet the fans' needs, and Epstein made a deal with Sean O'Mahoney to make his *Beatles Book* the official organ of the club (thus cutting down on the expenses of issuing newsletters). The great number of photographs and other material the clubs were distributing amounted to a large financial commitment for NEMS, but Epstein continued to think up new ways to communicate with the fans.

One of them was a cheap, thin plastic disc, the size of a 45 rpm single but played at 33 rpm, that could easily be sent through the mail and had been used successfully as an insert in magazines. This flexi disc (or flimsy) was used to send exclusive Beatles material to the fans each Christmas: messages full of jokes and puns, some songs, and that distinctive Beatles wit. Delivered to each of the twenty-five thousand fans in a bright yellow sleeve in 1963, it arrived as "Your own special Christmas gift." This was an innovative way to present the particular charm of the Beatles as a group, laughing, joking, and making fun of each other just as they did on the radio, but a flexi disc was much more personal. It was something you could own and show to your friends: here they were, the Fab Four, up close and speaking directly to you.

The NEMS management found many ways to keep the fans involved with the four musicians even as the Beatles were moving further and further away from them, finally isolating themselves and their besieged families from the hordes who had changed Beatlemania into barbarism. The four might be in hiding, but the clubs maintained a high volume of postal communication in which the fans were asked their opinion on important matters, such as the songs the band might record in the future. Members of the fan clubs received notes, official programs, and photographs with "special handwritten messages" from the Beatles. During the American tours, while Neil and Mal forged Beatles' signatures on piles of photographs, Epstein made the boys available to fan clubs, to talk with them and sign autographs backstage or in their hotels.[14] The routine was to arrive at the hotel, hold a press conference, then meet with deputations of fans or with kids who had won a chance to meet them in a competition organized by a radio or TV station. One small act of kindness was enough to make the wildest dreams of a fan come true and provide an attractive feature story for a local newspaper or television station. A chance encounter with the Beatles made someone famous for a few minutes and burnished the band's image as nice, ordinary boys who loved their fans.

BELONGING AND BEING THERE

The people around the Beatles noticed an extraordinary bond with their army of fans. It was "a special link that traveled beyond idolatry and entered the world of one-to-one chemistry and communion," said Larry Kane after he traveled the United States with the band. John Lennon joked, "We're the same as you, you know, only we're rich."[15] This bond was not simply the consequence of their delightful personalities, for both their music and their image were tailored to this end.

The bonding began with the records. Like other pop music at the time, the Beatles' songs were shaped to appeal to teenage girls. A common feature of their first records is their direct address to the listening audience, especially the girls: "Love Me Do," "Please Please Me," "From Me to You," and finally "I Want to Hold Your Hand" all have direct emotional empathy with the listener. Some songs were about the musicians' relationship with the audience. "From Me to You" mirrors the fan mail sent to the Beatles by the sackload, and "P.S. I Love You" comes across as a love letter. They wrote "Thank You Girl" as a direct message to their fans.[16] The Beatles were constantly touring, and their songs contain many promises to return home, with lyrics like "when I get home" and "when I'm home / everything seems to be right." They acknowledged the importance of postal communication: "I'll write home every day / and I'll send all my loving to you."

The theme of being away and returning home to a loved one is a strong thread in Lennon and McCartney's early work. There is also affirmation that feelings have not changed over time. Kenneth Womack thinks their lyrics express romantic love as a powerful, life-altering experience that brings total commitment and joy, but also anxiety and uncertainty. They sing of a carefree, exuberant love, considering it, in Womack's words, "a low impact sport."[17] Their early repertoire contained several covers of love songs recorded by American girl groups, and the sensitivity and sexual vulnerability of songs like "You Really Got a Hold on Me" were transformed when the Beatles interpreted them. These were the vehicles to explore the loud/quiet, tough/soft dynamic that intrigued the girls and impressed the boys. They offered intimacy—"Listen, Do You Want to Know a Secret"—but it was asexual and nonthreatening.

The way in which Lennon and McCartney built their songs manipulated the emotional responses of the fans, with melodies constructed around dramatic highs and lows, punctuated by soaring octave leaps and crescendos. The music worked in tandem with the screaming of the fans, prompting them to participate in the performance by answering the musicians with noise of their own. As the song reached a high point, a lift of a guitar neck or a quick shake of a mop top was the signal to bring eruptions of screams. It was calculated and effective. As Amy B. told them in a letter, "The way you shake your head during a show—oh, that really kills me!"[18]

The barrage of television performances, newspaper articles, and radio play that nourished Beatlemania put a premium on experiencing the band in person. It was the only way to consummate the relationship. The urge to get close to the Beatles overcame all considerations of propriety and acceptable behavior. It drove fourteen-year-olds to charge against burly police officers and commit small misdemeanors and dangerous feats, such as smuggling themselves in car trunks, dumbwaiters, and ventilation ducts. (The song "She Came in Through the Bathroom Window" described a real event when a fan broke into one of the Beatles' homes.) Once the miscreants had been corralled and led away, they always cited their love for the band as their motivation but usually could not give any better explanation of their objectives than wanting to get close to the Beatles.

The need to engage a physical dimension of Beatlemania opened up a lucrative market in relics of their American tour. The bed sheets they used, the carpet they walked on, and the stages they played on were often cut up into tiny pieces and sold to delirious fans. If the main point of Beatlemania was to get close to the band, it followed that you had to preserve that priceless experience. The need to own a little piece of the band—in the form of records, relics, photographs, or little bits of clothes and hair—motivated a lot of Beatlemania. A fan who paid to look around the trailer used by the Beatles before a concert said, after holding the towels and the glasses, "It made you feel close to him [Paul], that you got to touch something that you knew . . . they had touched."[19] The fans recorded and reenacted every part of Beatlemania in photographs, audiotapes, and artifacts of the experience.

Screaming started out as a form of applause, a participatory response that marked parts of the performance, but during Beatlemania it grew into a continuous wail that stopped at the end of the show (or in England, at the playing of the National Anthem). In fact it was never spontaneous; it was part of the show, and it could be turned on and off at will as the musicians on the stage provided the cues. The screaming had little to do with the music, but it was a critical part of the performance—a measure of the empathy between musicians and fans. People who went to the concerts and complained about not being able to hear the music were missing the point. The fans had bought the records and memorized the songs, so the reason to attend the concert was to be there and affirm their allegiance. The screams became a central part of experiencing the music in any of its forms. Girls wailed and shrieked relentlessly throughout the Ed Sullivan shows they watched at home and also during the showings of *A Hard Day's Night* in cinemas, not even stopping when the songs were over. At the time, the fans could come up with no better reason for screaming at the Beatles than their cuteness and the great music they made. Yet the sexual component of Beatlemania was recognized by the time the girls reached adulthood in the 1970s. "Years later I was thinking, well now I know why everybody was screaming, it was all this sexual energy, but at the time I was stupid or naïve, or too young to understand."[20]

Although NEMS management built up the feelings of longing and adulation, they cannot be blamed for inciting the frenzy. The fans themselves referred to this as "going nuts" and "carrying on," and this meant screaming, weeping, and desperate, violent rushes against walls of brick, glass, and people. They remembered it as a spontaneous, uncontrollable outburst: "I cried. I remember just sitting there crying. I don't know why," "I just screamed, I could not help it. It was like I had no control over myself whatsoever." Even staid, reserved adults could get carried away with the emotion. When George Martin and his companion, Judy Lockhart-Smith, went to a concert, "Judy and I just found ourselves standing up and screaming along with the rest."[21] Screaming might be an emotional response to the excitement and energy of a live performance, but it was nevertheless a learned behavior. Watching the audience on *The Ed Sullivan Show* was an education for the fans because it showed them what was expected of them. *A Hard Day's Night* was also a primer in Beatlemania. After seeing other people doing it, you followed suit when you got to the concert: "I want to scream but I don't know how . . . On the third attempt, I take in a big breath and emit brief squeals . . . The next one is perfect: a long, eardrum-piercing shriek. The girls in front grin and nod their approval."[22]

GIRLS JUST WANT TO HAVE FUN

It does not take a Freudian psychologist to suspect a sexual underpinning to the Beatlemania phenomenon, with its hysterical young women generating all that emotional energy. Many of the psychologists and sociologists called in by the popular press to explain Beatlemania stressed the sexual attraction between fans and the band. They reasoned that with all the stresses and strains that accompanied the spurts of rapid physical and emotional growth in the teenage years, there was a need for expression and release: boys had sports and gang violence (and later garage bands) as an outlet, but girls were much more restricted in what adult society would let them do. Beatlemania gave them the opportunity to reverse their roles and become more assertive.[23]

The girls in the audience were reaching that age when it was okay to have romantic (or even sexual) yearnings but were not quite old enough to do anything about it. A fan pointed out that sexual maturity came later in the 1960s than it did in the 1990s: "We weren't really into boys or anything like that. And all of a sudden these four guys come around with their charm, their music, their witty remarks, and it just kind of hit us like a ton of bricks." Another said "Boys hadn't really come into the picture at that point . . . and it was also like having a first love." And for the most part it was a pretty platonic first love: "This was never an overtly sexual thing, it was more like you want to love and kiss them." When reporter Gloria Steinem asked a tearful girl who claimed to be "passionately in love with Ringo" if she would actually go on a date with him, "she looked startled. I don't think so, I hardly know him."[24]

Thousands of girls made quite serious written proposals to the Beatles, offering love, marriage, and companionship. Karen R. wrote to John with tears in her eyes: "I could never love or marry anyone else. Loving you is all I do." Alice Z. sent this proposal to George: "Marry me! I promise I won't be a drag." Amy R. wrote to Paul, "I love you. I try so hard to show my love but what can I do? You are there and I am here. So I do my best. I have fainted for you six times." These girls did not scream at the Beatles, they screamed *for* them; it was the means to convey their love.

The links between Beatlemania and the development of the feminist movement in the United States are clear, and many scholars have interpreted Beatle-

mania as a significant uprising of women. Elizabeth Hess dates it as the beginning of "the second wave of the women's liberation movement" (the first had
begun with Betty Friedan and *The Feminine Mystique*). She remembered, "I was
just twelve, just beginning to understand that sex was power: my first feminist
epiphany." There was more than an undercurrent of rebellion in Beatlemania,
as thousands of middle-class girls took time off from being nice and instead
assaulted policemen or slammed their heads against glass doors. The girls who
ran amok in London and New York were part of that generation of "good girls"
who did not have sex before marriage, the reserved girls who did not get loud
and excited in public.[25] At the core of Beatlemania was a rejection of these role
models.

In their perceptive analysis of Beatlemania (cleverly subtitled "Girls Just Want
to Have Fun" after the joyous Cyndi Lauper song of the 1980s), Barbara Ehrenreich, Elizabeth Hess, and Gloria Jacobs conclude that Beatlemania was "the
first and most dramatic uprising of women's sexual revolution." They outline the
very narrow role models for "nice" teenage girls and stress the pressures forcing
them into marriage with a "nice" boy and then motherhood. This single-minded
pursuit of security and societal acceptance was temporarily disrupted by the introduction of a romantic but distant love affair with a film or rock star: "Adulation
of the male star was a way of expressing sexual yearnings that would normally
be pressed into the service of popularity or simply repressed."[26] Instead of being
courted and pursued by the male, the female Beatle fan could turn passivity into
aggression, actively and loudly pursuing the object of her sexual fantasies rather
than vice versa.

In his travels with the Beatles across America in 1964 and 1965, Larry Kane
got to know the fans pretty well. He soon grew to recognize those he calls the
"obsessed," the truly fanatical who thought they were either married to one of the
four or (in the case of the males) related to them or somehow deserving to be in
their presence. The obsessed were quieter and far more determined to get close
to the Beatles. When he confronted one young lady backstage, she told him, "Paul
is waiting for me. I'm late. He'll want to know where I am," and Kane was sure
she really believed this.[27] Geoff Emerick described the groups of sobbing teenage
girls who broke into Abbey Road studios as out of control, "the grim determination on their faces, punctuated by squalls of animal-like screaming." (He also
thought they made the energy levels high during the recording of "She Loves
You," which produced a better record.)[28]

In his intriguing re-interpretation of the hostile reaction to the Beatles in
America's Deep South, Devin McKinney argues that the band's critics coveted

the teenage girl who adored the Beatles and were jealous of "her fanatical willingness, initially, to follow, to be led and influenced; and ultimately her eagerness to act on the feelings that had been stirred inside her . . . She was impassioned by the Beatles where she wasn't by the church, or increasingly by the securities of bourgeois life."[29] The girls' screams for the Beatles might have been the first siren calls of a monumental event in twentieth-century culture, in which gender roles and the possibilities available to women were expanded and redefined. This is one of the most important of all the revolutionary changes of the 1960s because the changing status of women was felt all over the world, not just in affluent middle-class America. Scholars of feminism have labeled these changes as revolutionary, and no doubt the fans felt them deeply and personally. One said after the Beatles touched down at an airport, "When I arrived here I was normal. Now I'm all ripped up."[30]

At the height of their mop-topped adorableness, the Beatles stood for more than sex appeal, for embedded in their music and the stories of their dramatic rise to fame were other attractions: "I liked their independence and wanted those things for myself . . . We were so stifled . . . If only I could act that way, and be strong, sexy and doing what you want." "I didn't want to grow up and be a wife and it seemed to me that the Beatles had the kind of freedom I wanted."[31] The promise of freedom comes up often in the recollections of the fans, a freedom that reflected the Beatles' look, as well as their lifestyles and careers. The boys and the girls who listened to the music experienced Beatlemania as a liberation, an empowerment: "I tasted something. I was totally going outside of myself—it was total freedom. Once you tasted it you had to have more. The way you lost yourself in the crowd."[32]

You could not be part of this crowd without actively consuming the products of Beatlemania. You had to own the records and all the merchandise to be part of this group. Beatlemania involved a series of engagements and expenditures. It wasn't just boys buying the collarless jackets and Beatle boots—girls also wanted the clothes the Beatles wore: "I had to go out and get the John Lennon hat."[33] The Beatles inspired thousands of boys to form bands, but girls wanted to get into the act too. They did not rush out and buy guitars and drums, because these were boys' toys, but they could still imitate the band, pretending to be their heroes by singing the songs at private parties and events sponsored by radio stations.

Being like the Beatles was part of the process of getting close to them. Girls grew their hair and combed their bangs down across their foreheads just like the Fab Four. They organized themselves in groups of four and imagined which Beatle they represented. Each of them had a favorite, but they also spent time thinking about which one they might resemble. This worked for boys too. All of them added English or Liverpudlian slang to their vocabulary. Girls impersonated the band in private rather in the public gaze of the garage band movement and formed "tribute" groups in the privacy of their bedrooms. They acted out Beatlemania by coming together and lip-synching to Beatles records. They memorized all the lyrics and could even recite the dialogue from *A Hard Day's Night*: "My friends and I would act out the movie in our backyard. We got quite good at it."[34]

Beatlemania brought feelings of belonging as well as longing to the fans: "We were . . . a community, almost a family." Love of the Beatles managed to cross class and cultural boundaries, especially within the rigid social hierarchies established in high schools. It "cemented friendships with girls you normally would not have known."[35] The rituals of waiting in line, sharing favorite photographs, trading buttons or magazines, and then finally screaming at concerts brought teenagers together. Collecting Beatle memorabilia played a part in what Geoffrey O'Brien calls "the presumption of intimacy" that came after memorizing the records and the films, collecting the photographs and artifacts, and devouring every piece of information about the band.[36] Consuming all these products bolstered feelings of belonging and in some cases entitlement.

Beatlemania empowered the fans to identify themselves as a group. Love of the Beatles brought them together and helped articulate who they were. Many said later that the Beatles were their whole world at this time—a world that was divided between those who loved the band and those who did not. For those who had been transformed by the music, there was a sense of fellowship: "Everybody felt like they knew each other. You were immediately connected with all these kids. Everybody was instant friends." Over time this feeling of belonging assumed global proportions: "In the sixties they gave young people all over the world the sense that they were part of one culture."[37] Beatlemania fostered belonging, and screaming was the essential part of the behavior that marked admittance to the group. It was a rite of passage, the beginning of something transformative: "It was the first time I had ever screamed in public, and I was both surprised at my reaction and a little embarrassed at my outburst."[38]

The Beatles' tours of the United States in 1964 and 1965 meant much more than a series of lightning short performances accompanied by screams; they brought both joy and validation to millions and cemented the virtual bonds that had grown through mutual interaction with records, merchandise, and the media. To become a Beatles fan, as Geoffrey O'Brien recalled, was to enter the realm of "ferocious energies. Spectatorship here became participation. They were no longer to be any bystanders, only sharers." Poet Wyn Cooper recalled that becoming a fan made him feel part of "a larger community of people whose lives had Beatle soundtracks, haircuts and even a particular way of walking down the street."[39]

At the same time Beatlemania molded thousands of young people into a group, the Beatles had a marvelous ability to make individuals in that seething mass of humanity feel a direct, personal bond with them. Over and over again the fans remembered Beatlemania in terms of a personal relationship: "At the time it was something intensely personal for me and, I guess, a million other girls." When fans saw their favorite Beatle on stage, their hearts "missed a beat," and they imagined a powerful connection: "I was sure that Paul was singing directly to me."[40] Those girls who screamed and fainted shared common emotions and behavior, yet each one felt a personal relationship with her idols. Many experienced empowerment and a sense of destiny as they listened to the Beatles' music in the company of thousands of other fans. Francine B. wrote to them about their upcoming show at Forest Hills, New York: "You'll recognize me. I'll be in the 1st row in a yellow dress. By the time summer comes my hair will be very long. I will be holding a sign that says 'I'M THE ONE.'"

CHAPTER 9

CONVERGENCE

When asked to explain how his band had captured the hearts and minds of the global pop music audience, Brian Epstein said, "The Beatles are famous because they are good, but they are a cult because they are lucky . . . they have an extraordinary ability to satisfy a certain hunger in the country."[1] The Beatles happened to be at the right place at the right time, not just at that magic moment when they appeared on *The Ed Sullivan Show*, but throughout their long career. Had they emerged as a competent rock band a few years earlier or later, Beatlemania might have never happened because the special conditions that brought it to life would not have been present. They exploited a window of opportunity in the development of entertainment technology and transatlantic business networks that allowed them to saturate the media.

The Beatles arrived as a finished product at just the right time in the development of mechanized entertainment. They emerged when radio was king in popular music and left the world stage when entertainment was dominated by television. The deep roots of Beatlemania were growing well before the 1960s, when the advanced technology and manufacturing techniques developed during World War II were applied to entertaining the baby boomers who were born in the first years of peace. Social scientists and biographers have come up with some interesting and often imaginative explanations of Beatlemania, but its real foundation was technological rather than psychological. New technology was affecting all parts of the entertainment business, and the fruition of these technologies coincided with the demographic movements of the postwar boom. You don't really need a degree in sociology or psychology to understand why the Beatles conquered America in 1964; all you have to do is understand how all the new machines worked in their favor, and then do the baby boom math.

The Beatles rode the waves of technological innovation and demographic trends across the Atlantic, picking up these trends as they formed in the UK and joining the great surges of change occurring in the United States. As John Lennon said: "Whatever wind was blowing at the time moved the Beatles, too. I'm not saying we weren't flags on top of a ship; but the whole boat was moving."[2] The Beatles and their management had the intellectual curiosity to identify the significant trends in technology and culture and then exploit them. And their timing was always impeccable.

DEMOGRAPHICS

The driving force of Beatlemania was the growth of the audience for popular music. The great increase in the birth rate immediately after World War II, the baby boom, has shaped American and British popular culture for half of the twentieth century. The baby boom was a demographic cohort of around 77 million people born between 1946 and 1964, and this huge audience took the Beatles as their own. The number of Americans between eighteen and twenty-four increased by almost 50 percent during the 1960s, from 16.5 to 24.7 million. By 1964, about 40 percent of all Americans were aged twenty or less, and seventeen-year-olds were the largest age cohort in the United States, a lucrative market for a range of consumer goods and services, including records: 11 million girls had bought a good half of pop records in America by the time the Beatles arrived.[3] Teenagers in the UK were not as numerous or as affluent, but they still had an impressive spending power. In the 1960s about 7.5 million of them were between twelve and twenty-four, and they spent about 3 million pounds a day. Out of an average take-home wage of five to seven pounds, the typical teenager would spend at least two pounds on clothes and buy at least one record.[4] English teenagers did not have the cars or the televisions of their American counterparts, but as the economy grew steadily in the 1960s, they were able to buy transistor radios, portable record players, and tape recorders. What made the baby boom generation different from all those who went before was its affluence.

Demographics can explain the Beatles' huge sales figures, such as the 2.5 million records they sold in the United States during the first weeks of 1964. These were well-made records, but were they that good? Or was it that millions more people were now buying records? American record companies had already seen it coming. In 1954 they estimated that 5 million new record buyers would enter the market by the end of the decade. They were right. The sales of records doubled during the 1950s. By 1970 the average teenager was spending at least five times as much on records as he or she did in 1950.[5] In England there could be no doubt that more people were buying records; the 9 million pounds spent on records in 1955 had increased to 14 million in 1957 (the result of the skiffle boom) and to 22 million by 1963. In that year the *Daily Mirror* was publishing stories about the 5.5 million "spendagers" in the UK, who were buying around 50 million records a year. Each week these consumers bought over 20 million pounds' worth of clothes, tape recorders, record players, and records. Every week they bought nine hundred thousand copies of teen magazines like *Fabulous*.[6] Record sales reached a peak in the UK just as the Beatles hit their stride as studio musicians.

By the early 1960s the Beatles could appeal to the largest and most affluent group of teenagers there had ever been. Unlike other bands of their era that captured a portion of the youth market but could not hold it for very long, the Beatles held onto their fans for nearly a decade and along the way gathered more of them from broader age cohorts. As children of the war years, the Beatles were a few years older than the first of the baby boomers, and this slight age difference made them ideal heroes—older but not too old to push themselves out of the peer group into inaccessible adulthood. They kept pace with the spending power of this generation, who moved from buying singles to albums and from spending a few shillings to get into the Cavern to paying pounds to watch the Beatles in movie theaters. The Beatles were not the first entertainers to expand their fan base to embrace larger and older audiences—this was the path of pop music success that Brian Epstein had mapped out for all his groups. But while the Beatles engaged the adult audiences who tuned in every week to *Sunday Night at the London Palladium* or *The Ed Sullivan Show*, they were attracting the attention of younger and younger fans. The Beatles appeared at the right time with the right products in the flowering of baby boomers as consumers and, most important, managed to hang on to them longer than any other pop group.

Their music, look, and humor appealed to all ages. There was something in the Beatles' music that connected to the very young. After their triumphant tours of the United States, they appeared in a cartoon every Saturday morning from September 1965 until April 1969, and these were later shown in the United Kingdom and all over the world.[7] The cartoons were rebroadcast in the 1970s as the continual discussion of a Beatles reunion kept the band in the public eye, and their music was still heard everywhere. The teens who bought Beatles records in the 1980s and 1990s had started out watching the band as cartoon characters in the 1970s.

Much has been written about the development of their music as the Beatles transformed themselves from teen idols into symbols of the sixties counterculture. They moved from producing single after single aimed at the pop audience to exquisitely crafted albums that appealed to highbrows and music critics. This transition was as fast as their rise to global prominence, only taking two years to move from "She Loves You" to "In My Life," and then another two from *Rubber Soul* to *Sgt Pepper*. The Beatles were able to engage older listeners and broaden their appeal to both sexes over time. Their growing artistry also attracted more males to their fan base, because boys were usually more interested in the music than the image, and many of them hoped to form a band just like the Beatles.

The progression of the band's music went hand in hand with the development of the pop music record as a consumer product. The industry they entered in the early 1960s was dominated by the single, which at 59 to 99 cents in the United States and 6 shillings, 3 pence, in England was well within the financial resources of young consumers. Many millions of babysitting and yardwork dollars and pounds were invested in Beatles records. In England the sales of singles reached a peak in 1964, accounting for eight out of ten records purchased, just at the right time to turn the Beatles into pop stars. But from then onward the proportion of singles in total sales decreased and the long-playing record became the vehicle for popular music, just as the Beatles moved from recording singles to concept albums.

Their involvement in films helped them engage a wider customer demographic. Their first film was directly aimed at the American teen audience who were desperate to see their heroes for the first time or eager to relive their concert experience—young girls stood in line for the film and watched it repeatedly. Yet the film also received acclaim from the critics and the adult audience, who might have been embarrassed by Beatlemania but were impressed with the whimsical humor and avant-garde pretensions on film. The transition into serious artistes certainly lost them a portion of the teen and preteen audience, who were often confused by their post–*Rubber Soul* music and soon migrated to other adorable mop-top bands like the Monkees, but the Beatles as film stars still managed to keep a foot in both camps. The chief artifact of this accomplishment is their cartoon *Yellow Submarine* (1968), a full-length feature that impressed film and music critics with its inventiveness, yet captivated the very young, who consumed its pictures and music just like all the other cartoons they watched. Good cartoons are shown for decades, the best are immortal, and the Beatles' cartoons played an important role in introducing them to each succeeding generation.

Films brought massive new audiences to experience Beatlemania. It is hard to disagree with Bob Neaverson's conclusion: "Without film, the Beatles' global popularity would not and could not have existed to anything like the same degree."[8] The conflation of concert tours, records, and films into one text of Beatlemania was Brian Epstein's most brilliant idea. It exploited the synergy of different entertainments and technologies while creating a media narrative that was greater than the sum of its parts.

The army of salespeople and manufacturers who got onto the Beatles' bandwagon played their part in the media frenzy. Although popular music was already recognized as a means to reach young consumers, it took the Beatles to reveal its marketing potency. As soon as television producers like Jack Good and Dick Clark married the music to the image, pop music became the most effective vehicle to sell stuff to teenagers. Television and films exhibited fashions and the latest dances as well as the performers. They showcased a wide variety of goods and services. Beatlemania promoted the output of the multinationals that produced the records and the portable players, the agribusinesses who made chewing gum and sweet carbonated drinks, and the chemical companies who made toothpaste, dandruff shampoo, and acne creams. The generation that put Beatlemania on the map were throwing away their 1950s clothes and hair products, and the Beatles as usual were just ahead of them, as they moved from being Teddy Boys to England's most famous hippies. Embedded in all those photographs of girls screaming are the signs of significant changes in teen fashion, from the Bermuda shorts and bouffant hair styles of the early concerts to the full psychedelic regalia of the later shows. The Beatles were in the right place at the right time with the right music and the right clothes.

An article in *Newsweek* in 1963 added up the value of all the products sold by the band and found that merchandising far outstripped the sales of records and concert tickets. In 1964 the Beatles were expected to sell about 50 million dollars of merchandise. The company licensed to sell Beatle wigs was turning out about fifteen thousand a day, and many more unlicensed manufacturers were making this popular item.[9] The Beatles' image was put on lunchboxes, sweatshirts, table lamps, and combs. The fans were collectors who traded badges, pins, photos, and even cards from Beatles bubble gum, "that rotten gum we had to chew for those pictures in the package." At the height of Beatlemania one of the band's sponsors rightly judged the Beatles to be "the most powerful salesmen in the world today."[10]

TIMING

In November 1963 Brian Epstein made a preliminary trip to New York to finalize preparations for the upcoming tour, conferring with agents and promoters, visiting equipment manufacturers, making contact with the record companies who were handling his groups in the United States, and buying clothes for himself

This picture illustrates some of the core elements of Beatlemania: happiness, together-
ness, and merchandising. Complete with Beatles T-shirts, these two girls wait behind a
police barricade to get a glimpse of their heroes. Women dressed alike might have typically avoided one another at social events, but for Beatles fans, wearing the same outfits
marked their membership in a dynamic group. It signified their identities as Beatles
fans. (Courtesy National Museums Liverpool)

and his travel companion and protégé, Billy J. Kramer. Here are all the different
elements in Epstein's management of the Beatles, from making records to de-
signing the look, and here we see the careful preparations for his masterstroke.
In an interview with the *Daily Mirror*, Epstein revealed that "every aspect of their
lives is carefully planned . . . planning and timing are desperately important."[11]
Billy J. Kramer remembered that Epstein and the potential of the Beatles did not

make much of an impression on the record company and venue management executives they visited, but Epstein got what he came for with top billing on *The Ed Sullivan Show*. During the visit he constantly asked American businessmen questions about the media, rather than the music.[12] He had come in person to New York to find out all he could about American radio and television, for these were the tools he would employ in the band's upcoming tour.

Beatlemania had begun at a fortuitous moment in the development of British television and radio. The BBC monopoly of pop music broadcasting (with the distant and often faint competition from Radio Luxembourg) meant that if it did embrace a band, the group would dominate the airwaves. Because of the lack of competition, a program like *Saturday Club* had a captive audience of more than 3 million music-mad teenagers, who made up the bulk of English record buyers.[13] Had pirate radio arrived a little earlier to challenge the BBC monopoly and steal away some of its young listeners, or had the Beatles peaked a little later, this huge listening audience would have been lost. The demand for rock and pop music programming could not be met by the BBC, which opened the door for independent radio stations that operated outside the law.

The first pirate radio station was Radio Caroline, which started broadcasting in 1964 from a vessel outside the UK's three-mile boundary. Because its programming depended totally on records, pirate radio delivered a far broader spectrum of popular music than the BBC, and unlike the state broadcaster, it played only pop. The pirates saw no need to pay royalties to record companies or artists and were not tied to them as Radio Luxembourg was. Other stations soon joined Caroline, helping out independent companies by making many of their records hits at the expense of the majors like EMI. After the government closed down the pirates in 1968, the music weekly *Disc* lamented: "Expect many of the small independent record companies to fold now there is no pirate radio."[14]

The situation of American radio broadcasting when the Beatles first arrived in North America was completely different but no less advantageous. Instead of monopoly there was diversity, but the competition of television was driving important changes in radio organization and programming. The rise of television as a mass entertainer forced radio and film to look for new customers, because families now gathered around their television every night. Radio still reached a national audience through linked networks like NBC and CBS, but individual radio stations had to aim their programming at a local audience.

Instead of the free-for-all of the 1950s, when stations played a broad mix of popular, classical, country, and religious programming, and used live musicians as well as recordings, radio in the early 1960s was settling into the Top 40 format, which depended on recorded music. The idea for this format came out of jukebox operations, where market research had shown that customers tended to play the same songs over and over again. In Top 40, the national charts determined what was being played, and these charts reflected sales in the big urban areas, which were easiest for the record companies to register and manipulate. Although Top 40 put a lot of power in the hands of program directors (who assembled the playlists), radio's celebrity deejays were still very powerful. In the big urban markets, the high-powered AM stations represented the networks, but the rise of independent stations provided more competition and put a premium on hiring a high-profile deejay and being the first to exploit the latest hit.

The radio industry the Beatles first engaged in North America was highly competitive, with long-running rivalries in most of the major cities they visited. In New York WMCA, WINS, and WABC struggled to dominate the biggest and most influential radio market in the country. In Cleveland it was WHK versus KYW. In Miami WFUN and WQAM battled it out. Top 40 radio was drawn to the Beatles because its programming was linked to the charts, and Beatles records dominated the Top 5s and 10s. This symbiotic relationship helped the band monopolize the airwaves as well as the charts. The big stations in urban markets soon figured out that the Beatles were the key to getting an edge on their rivals. Each of them played the records nonstop and claimed the strongest links with the band. Thus WFUN became "Your Beatles station in Miami," and so on. The Beatles' management was also aware of this critical relationship: "They knew that there were hundreds of Top 40 stations and they were the keys to everything," said a deejay.[15] Stations had to do more than just play Beatles records; they had to demonstrate an exclusive connection with them. This explains the frenzied response of the New York radio stations to the Beatles when they arrived, and the legions of deejays and radio correspondents who followed the band as they toured America. Each town they visited was just another battlefield in the radio wars, and radio stations competed strenuously to put on Beatles concerts, obtain advance release copies of the records, and broadcast interviews with the band members.

Radio provided the most immediate link between the Beatles and their followers, disseminating information, news reports, and special interviews that brought the band into the lives of radio listeners at a time when deejays were active in their communities and the fans viewed these on-air personalities as their friends. Local radio stations encouraged requests, organized fan clubs and concerts, compiled their own local charts, and devised daily contests to bring fans to the station.

These were not small-scale efforts; for example, a WMCA wig contest got eighty-six thousand entries.[16] If anyone conducted or orchestrated Beatlemania, it was the radio deejays rather than the Beatles' management. The deejays maintained a continual dialogue with the fans over the air. They stoked the excitement, kept up anticipation with scraps of Beatles news, and were cheerleaders of concert frenzy. The Beatles helped out a lot of deejays, including characters like Murray the K, who high-jacked media attention in their first tour, but in return they won over the shock troops who would lead Beatlemania from the front.

TELEVISION

A few steps removed from the raucous radio announcers were the local television stations, who kept a respectable distance from the excesses of Beatlemania but nevertheless looked to it for programming. The interview with a local deejay, or a picture taken by a station publicist, would often find its way onto local television and then onto the national networks. Television not only exploited the entertainment value of popular music, but also saw its potential as news. The delivery of the news on television was changing like everything else. The five-minute news briefs of the late 1950s were expanded to all of fifteen minutes in the early sixties, and in the tumultuous year of 1963, CBS took the big step of increasing the program to thirty minutes and making the news anchor, a former press journalist named Walter Cronkite, the center of the show. This was an eventful year and a massive boost for television news. The Cuban Missile Crisis, civil rights demonstrations, and the tragic death of the president made television the prime disseminator of news. In the words of Erik Barnouw, those four days in November were "the most moving spectacular ever broadcast," and they brought television news to the pinnacle of its influence.[17] By the time the Beatles touched down at Kennedy Airport, a majority of Americans got their news on television, and Walter Cronkite was the most trusted man in America. Television news introduced the Beatles to America and continued to promote Beatlemania with its growing technical expertise in remote broadcasts from the field.

The development of the televised variety show also came at the right time for the Beatles, as they moved from a scruffy beat group to all-around entertainers. In another example of their perfect timing, they emerged as polished television performers just as the medium was reaching a mass audience. Their appearances on *Sunday Night at the London Palladium* on one side of the Atlantic and *The Ed Sullivan Show* on the other occurred when these programs were attracting their peak audience, including millions of people who would not have bothered watching a pop group. A few years later, specialized pop music shows aimed solely at teenagers had siphoned off much of this audience.

While Jack Good was pioneering teen music shows in Great Britain, *American Bandstand* was becoming an institution of American youth culture. It began in 1957 as the Philadelphia-based *Bandstand* and was soon broadcast to a national audience. Under the direction of Dick Clark, it became a powerful marketing tool for the records mimed on the program, although it had little of the excitement or the immediacy of the English teen music programs.[18] Beatlemania encouraged the production of many more television shows aimed at the teenage audience. In June 1963 *Thank Your Lucky Stars* broadcast a "Mersey Beat Special" that featured many of Brian Epstein's groups and garnered an audience of 6 million. It was the work of Philip Jones, who had joined Granada TV from Radio Luxembourg and was one of the first broadcasters to see the potential of the Beatles when he first booked them in January 1963. By 1964 television stations around the world were scrambling to get some Mersey Beat for their programs, and American networks were frantically courting Jack Good. ABC debuted *Shindig* in September with a taped version of an English *Ready Steady Go!* program that was edited for American viewers. Good continued to produce shows for American television that featured the Beatles and other British Invasion acts, including a special program called *Around the Beatles*, which was broadcast on both sides of the Atlantic.

The visual impact of the British Invasion established youth music as a permanent fixture on television rather than a small daily dose or a weekly review show, and by the mid-1960s much more programming, like *Ready Steady Go* and *Shindig*, were competing for the attention of the teenage viewer. If the Beatles had emerged at this time, their television exposure would have reached a much smaller and more narrowly focused audience than those they engaged on the weekly variety shows a few years earlier. Would their impact have been so powerful if they had appeared on *American Bandstand* one summer afternoon instead of on *The Ed Sullivan Show* that frigid Sunday night, when the whole nation was sitting in front of the television? Probably not.

As soon as the Beatles made their successful debut on national television, their management ran down their radio exposure. After they returned from their first American tour, Epstein would only release them to make special programming, which was broadcast a few times each year during national holidays like Christmas. The same policy was applied to their television appearances. In the beginning of their television careers, the Beatles were happy to appear on variety, children's, and comedy shows, but as their fame grew, they became more discerning, moving to the same special event strategy that had worked so well on radio. The Beatles made thirty-seven appearances on television in 1963, but by 1966 they were down to four.[19] Tape-recording technology made all this possible, because it ensured that every minute of the Beatles' output could be rearranged, recycled, and rebroadcast to fit the extraordinarily crowded schedule of the band. Their management always adopted new technology with a view to exploit the Beatles' time to the fullest.

Under Epstein's direction the Beatles became as skilled at television performance as their management was at negotiating special programming for the band. They appeared on scores of television shows between 1963 and 1965, becoming as ubiquitous on the small screen as they had been on radio. Their triumph at the *Royal Variety Performance* in November 1963 was not a lucky accident; it was the result of months of preparation and experimentation in getting the sound and look across perfectly. By the time they went to the CBS television studios on 53rd Street in New York to do *The Ed Sullivan Show*, they were masters of the media. No other pop group had as much experience on television as the Beatles. The day after they arrived in New York, they went to the television studios for a rehearsal and sound check. The technicians at CBS were impressed by the time and care the musicians and their road crew took in their extensive rehearsing, in which they practiced the songs and their positions on stage. After a thorough sound check, they carefully marked the levels of the mixes on the control console. The CBS crew was quite surprised when the group asked to listen to a playback of their rehearsal—nobody had ever asked for that before.[20]

The Beatles had the technicalities all worked out and went through their preparations like the seasoned television performers they were. On the Sunday afternoon before the show, they went back to the studio and performed a complete run-through of all the songs before a specially invited studio audience. A different audience was seated for the live performance at 8 p.m. *A Hard Day's Night* gives us a detailed view of the daily work of the band in 1964, with lots of songs, news conferences, fittings with tailors, and a detailed narrative about the production of a television show that showed all the equipment involved. There is no scene of them working in a recording studio. At the height of Beatlemania, making appearances on television was as important, if not more important, than making records.

As the Beatles' annus mirabilis took shape, each step of the way was marked by a televised performance. Their appearance on *Thank Your Lucky Stars*, the top pop music show in England, was probably the critical element in pushing them and "Please Please Me" onto the national stage. The fact that they got this op-

portunity showed that they were now connected with the big players in the entertainment business. It also showed how important it was to have a management
organization in London. Brian Epstein called on Dick James in his offices on
Demark Street to negotiate the rights to publish the Beatles' music. James picked
up the phone and got the band on the show to demonstrate his influence. Epstein
gave him the rights and made him a very rich man, but the Beatles got a priceless
boost of publicity and prestige in return—just when they needed it most.[21]

Television was the most important ingredient in the bubbling stew of machines, media, and mass entertainment that made Beatlemania possible. The
rise of the Beatles occurred just as television was connecting to a new audience
of young music lovers. For all the weight that the guitars and amplifiers had on
the Beatles sound, and for all the importance of tape recording in making their
records, the most valuable technology in the Beatles' arsenal was television. Managing this media exposure was the secret to their phenomenal rise. They were the
first band to negotiate the changes wrought to popular music by television.

THE STORY THE PRESS WAS WAITING FOR

The Beatles as a news story also developed at exactly the right time. In the 1960s
both the American and the English newspapers were putting emphasis on feature stories about personalities when many of these new celebrities came from
the world of entertainment. Competition with television news forced the print
media to invigorate their content with more images and to appeal to a younger
readership. People were getting their news in visual form before they went to bed
rather than first thing in the morning, when the paper was delivered, and thus
newspapers countered by offering more feature stories and putting more photographs on their pages.[22] Television programmers were also looking to make their
news (and newsreaders) more attractive and to personalize the news with stories
built around interesting people.

The Beatles came along when the tabloids were at the peak of their powers in
the United Kingdom. In 1964 the *Daily Mirror* sold about 5 million copies a day,
giving it a readership of around 12 million Britons, one-third of all the adults in
the country.[23] The tabloids were moving eagerly into feature stories about youth
music, mod fashions, and teenage angst. The scale of Beatlemania and its juicy
connections with other tabloid obsessions—the new rich, teenage delinquency,
the swinging set, and the competition with the United States—made it the story
of the decade. Even after the Beatles had broken up, the press kept the story alive

by floating rumors about a reunion or the death of a band member. For tabloids like the *Daily Mirror*, Beatlemania was a very important story. Don Short, the *Mirror*'s entertainment writer, was close to the band and their circle; he was a journalist they could trust, in the words of Tony Bramwell. In exchange for adhering to the NEMS line, a few favored journalists like Short were supplied the exclusive material their editors craved above all else. This relationship helped Brian Epstein control the image he had created for the band.[24]

Running parallel to the rise of the feature story, and the creation of a new set of rock'n'roll celebrities, was the growth of music journalism. Entertainment in the 1950s did not have the place it now occupies in our newspapers and television programs. It was a sideline, tucked away in the back of the paper, a small part of the overall picture of the world dominated by great events and important people. The national papers did not have much time for pop musicians unless they were misbehaving. This lack of coverage pushed fans and amateur journalists into starting up periodicals devoted to the local music scene, like *Mersey Beat*, which would play such an important part in publicizing the Beatles. The Beatles emerged into public consciousness as more journalists moved into reporting about popular music and its even more popular musicians. Even the stuffy broadsheets like the *Daily Telegraph* developed an interest in the pop music business and joined the tabloids in publishing a weekly top ten of record sales.[25] Over two hundred journalists attended the Beatles' first news conference at Kennedy Airport. The people who attended later conferences at the Plaza Hotel said they were as big and as buzzing with excitement as the press conferences given at the White House by a young and extremely media savvy president.

The Beatles managed to engage the press on many different levels. The tabloids were always hungry for pictures and gossip, and the magazine trade also depended on a steady diet of interviews and photographs. Even serious newspapers like the London *Times* took an interest, and its music correspondent took time off from reviewing symphony orchestras to discuss the finer points of the Beatles' music. In the United States, the weekly news magazines *Time* and *Newsweek* broke the news first, and as soon as the band arrived at Kennedy Airport, the New York dailies rushed in, and the rest of the country's newspapers piled on. At the other end of the news hierarchy, local papers engaged budding young journalists to write columns about music and fashions inspired by the Beatles. These "teen editors" helped newspapers reach affluent young consumers and attract advertisers. A trickle-down effect of the British Invasion enriched businesses far removed from popular entertainment. It wasn't just about music; it was also about the accents and the clothes, and consequently retailers organized fashion shows around the new "London" styles and brought in amateur guitar bands to give the shows some swinging appeal.

The Beatles became famous when images were critical to news coverage and newspapers were putting illustrated magazines in their Sunday editions. The *Sunday Times* introduced a color magazine in 1962, employing a photogravure process that made the images clearer and brighter. The potency of the image was one of the important threads in the construction of the "Swinging Sixties." The Beatles' carefully manipulated look reached full bloom just when the press was craving attractive images: the suits, the hair, the guitars, and how they held them on stage all came together effortlessly into one irresistible whole at exactly the moment when the Beatles' abilities as pop musicians had reached a peak.

MEDIA SATURATION

The image, the music, and the media exposure came together in 1963. Whether by luck or by incredible planning, Brian Epstein and the management team exploited it all to saturate the media, first in England in 1963 and then in the United States in 1964. The strategy blanketed radio, television, and the press and worked to exclude other acts and interests. This phenomenon was first explored by the French writer Jacques Attali, who realized the power of repetitious acts in fundamentally altering codes of social reproduction. Recording technology turned performance (representation) into machine-made copies, and music into a "repetitive commodity," colonized and sanitized. Attali looked forward to a society in which "nothing will happen anymore."[26]

When Beatlemania first appeared on the front page of the tabloids in November 1963, it was "happening everywhere!" and by February the next year, the press was saying, "You can't get away from them." Even the wife of the British prime minister, Mary Wilson, felt she had to tell the public that "the Beatles are quite the favorites in our home."[27] The Beatles' brand became so powerful that it trumped all others: Capitol Records answered its phones as the Beatles' record company, Pan Am renamed the jet carrying the Beatles after the band, and numerous radio stations claimed to have been taken over by the Fab Four. This media saturation had a life of its own—generating interest from far away and then building up its energy as it came closer. It created a sense of anticipation that fueled the mass hysteria for the whole of 1964.

Media saturation meant that the binary code of 1960s print and broadcasting journalism split the world of popular music between the Beatles and all the rest. Along with their music on the radio and constant exposure in the press, the Beatles on television forged a media ubiquity that made it difficult to avoid them (if you were young and connected to the world of pop music—large segments of the population had no interest in them). During the exceptionally cold winter of 1962–1963, the more than usually lethargic British radio and television audience was bombarded with the Beatles. Other records had been overplayed on radio and jukeboxes, and there was even a 1963 pop song ("Footstomp" by Jet Harris and Tony Meehan) about this saturation: "Now I've heard that tune so much / I wished that I was dead." But the Beatles' music seemed to be everywhere, and there was no escaping it. The exposure to their records coincided with the massive press coverage that left no event in their lives unreported and made the individual Beatles seem closer and more familiar than any of their contemporaries.

You might know of the Dave Clark Five or Brian Poole and the Tremeloes (the band that Decca chose to sign rather than the Beatles), but you did not know the names (or the personalities) of the members of the band because there was nowhere near as much information about them. A lot of Americans (and several British newspaper editors) did not realize these two groups were from London, not Liverpool. The Beatles got so much airtime that you felt you knew them as individuals, recognizing their faces and voices, and building a character profile of each member. Did anyone know the name of the drummer of the Big Three or the Swinging Blue Jeans, the Liverpudlian bands who were competitors of the Beatles? But Ringo, on the other hand, was a name everyone knew.

The Beatles invasion of America had really begun months before they landed in New York. The sequence of press articles, television spots, radio play, record releases, and Capitol's massive marketing campaign prepared the way for the personal appearances that culminated this long period of incubation. Looking at this process from the viewpoint and timeline of an American teenager is instructive. The first inklings that the Beatles were something beyond the average boy band came in the televised reports of English Beatlemania and the press articles in November and December 1963. A very small number of Americans who had penpals in Liverpool have the honor of being the first American Beatle fans. Some of them even got the records mailed to them. Then came the radio play, muted at first, then growing in intensity as Capitol racked up the promotion for "I Want to Hold Your Hand." While everyone in high school was talking about "I Want to Hold Your Hand," the news broke that the Beatles were going to appear on *The Ed Sullivan Show*, and this ratcheted up the suspense a few more notches: "It just started that incredible curiosity . . . I couldn't wait to see these guys on tv."[28] If you missed that show, you were immediately told about it at school the next day, and then you had two more chances to see the band as they appeared on Ed Sullivan for the next two weekends. Meanwhile their records were playing continuously on the radio, as many stations switched all their programming to Beatlemania.

The Beatles' first visit to America was not really a tour, because they played only three live performances for audiences in Washington and New York after *The Ed Sullivan Show*. The real tour was scheduled to begin in August. Between the band's departure from New York on February 21 and their arrival back again in Los Angeles on August 18, Vee Jay Records released an EP of earlier material in March, Capitol released the *The Beatles' Second* LP in April, another EP in May, and an LP, *Something New*, in July. The large number of Beatles records in the April *Billboard* charts shows how quickly the record companies responded to the band's popularity by releasing as much music as possible. A film of one Washington performance was shown in selected theaters in March, and the *A Hard Day's Night* soundtrack album was released in June. In July came the news that the film would soon be released in the United States, which was a welcome break in the fans' "long wait" for the Beatles as the summer vacation wore on.

Cinemas and radio stations began promotional campaigns that made buying a ticket for the Beatles' film an event, and the lucky fans who had camped out all night got special badges that said "I've Got My Beatles Movie Ticket, Have You?" *A Hard Day's Night* was an exceptional film in many ways, but it worked best as a substitute for the live performances and sense of intimacy that was central to Beatlemania. The film opened simultaneously in over five hundred American cinemas in the summer of 1964—perfectly placed between two American tours. As another example of Beatlemania, it generated as much news coverage as the concerts. The blanket release was a complete success, and it would lead to the mass marketing of blockbuster movies in the 1970s and 1980s, in which intensive advertising was followed by a simultaneous opening in hundreds of theaters.

The film was marketed in tandem with the soundtrack album. After hundreds of thousands of fans had seen the film repeatedly over the summer, the excitement was in full swing by the time the band actually arrived in August to start the tour. Radio stations announced the dates for the tour as early as March, and fans lined up, with media attention on the lucky ones as they posed triumphantly with tickets in hand or on the unlucky who moaned, "I'll just die if I don't get one." Radio stations and concert venues were in full swing with competitions and other promotions linked to the acquisition of those priceless tickets.

The Beatles' management staged the same process of fostering anticipation and building a surge of multimedia support to promote *Sgt Pepper's Lonely Hearts Club Band*. The ten-month gap after the release of *Revolver* had led to rumors that the band was either running out of inspiration or engaged in a historic project that would change pop music forever. The release of "Strawberry Fields Forever" in January 1967 confirmed that the Beatles were indeed making extraordinarily ambitious pop music, which only increased anticipation of the album. Tapes of a song called "A Day in the Life Of" leaked out, and as Greil Marcus said, "The record, unheard, was everywhere."[29] When EMI and Capitol finally released it in June, tracks from the album swamped the airwaves in America and Europe. While Langdon Winner was writing that the release of this album was the closest Western civilization had come to unity since the Congress of Vienna in 1815, Greil Marcus was crossing the United States on Interstate 90, hearing it everywhere he went, "wafting in from some far-off transistor radio or portable hi-fi." Lester Bangs has pointed out that the British Invasion was more about an event than the music, but this event swept all before it, for as Paul McCartney admitted, there was a time in the 1960s "when everything was about The Beatles. We were simply everywhere you looked."[30]

CHAPTER 10

TECHNOLOGY

In many ways, the Beatles were children of World War II. All four of them were born during the war years, and John Lennon allegedly arrived during one of the German air raids on Liverpool. The technology that framed their world and facilitated their rise to international stardom was also born during the war. The global conflict produced a wave of innovation, including jet engines, computers, and the advanced electronics incorporated into radar and mass-produced televisions. It also affected entertainment in many ways. The armies fighting the war sent so many messages that the wavebands carrying radio and radar signals were overcrowded, which forced researchers to explore the very high frequencies (VHF-UHF), which would carry the new media that changed popular culture in the 1950s. On the bottom of the sea, experiments in sound detection and insulation would influence sound recording and transatlantic communication. In the air, space travel and jet-powered airplanes completed the technological conflation of time and space begun by the Victorian engineers who had laid railroad tracks and telegraph lines. The most expensive research program ever undertaken brought wartime innovation to a climax: the Manhattan Project began the atomic age with a bang. Military technologies were applied to every walk of life in the 1950s, from the Plexiglas that enclosed the cockpits of airplanes, which the surgeon Harold Ridley used in the first intraocular lens transplant, to the shape of these cockpits, which influenced the design of the wonderfully modern Wurlitzer jukeboxes.[1]

The rise of the Beatles as the world's biggest band coincided with the broad integration of wartime technology into consumer products. Their first trip across the Atlantic marked another milestone in the transportation revolution. They arrived in New York on a Pan Am 707-321B aircraft, named *Clipper Defiance*. Boeing had developed the 707 out of a military project, the KC-135 tanker, in the 1950s, when the idea of transatlantic passenger jet travel was far from anyone's mind. But Juan Trippe of Pan Am had a dream of fast, cheap jet travel, and he worked with Boeing to develop a passenger jet. Boeing introduced the 707-120 series in 1958, and Pan Am led the way in adopting them. Transatlantic travel became much faster and cheaper, and Boeing became a producer of passenger jets rather than war planes. The requirements of the Cold War, and the recent history of jet-to-jet combat over the skies of Korea, forced a rapid development of jet engines in the 1950s. Pratt and Whitney produced a much more powerful engine, the JT3D turbo fan, that was adopted by Boeing for its 707-320 series—larger, faster jet aircraft with true intercontinental capability. Had the Beatles flown on a 707-121, they would have landed in Gander, Newfoundland, for refueling, but the 707-321 could fly nonstop to New York and arrive in triumph.

The success of the Boeing 707 opened the skies for transatlantic air travel, which thirty years before was considered a very long and dangerous trip that only adventurers would risk. During the war it took over twenty hours to make the crossing in four-engine bombers like the Consolidated Liberator, but the jet liners could do the trip in about nine hours. The introduction of the Boeing 707 and the rise of the charter airlines, which meant much cheaper tickets, dramatically increased air travel between Europe and the United States. The Beatles led the way, but thousands would follow, including a generation of European students (that I was among) who suddenly found transatlantic travel and the wonders of America within their budgets.

After 1964 the trickle of passengers crossing the Atlantic became a flood, as ticket prices dropped and more carriers entered the market. In 1956 when rock'n'roll arrived on British shores, about 785,000 passengers had made the Atlantic trip. By the time the Beatles set off for America, the number had reached 3 million, and by the end of the decade, it was over 7 million. During the same period, the number of seats available on transatlantic flights went from 5 million to 13 million. With as many as one hundred thousand flights across the Atlantic every year, avoiding collisions and managing the airspace over the ocean became major issues, but these problems had been tackled during the war, and those technological and organizational solutions became the systems that kept transatlantic travel safe during the 1960s.[2]

The story of how "I Want to Hold Your Hand" was introduced to America shows how improved transportation and communication aided the diffusion of culture. Capitol set the release date for January 13, 1964. A teenager named Marsha Albert saw the CBS television piece on December 10, 1963, and wrote to her local Washington, D.C., station, WWDC, asking them to play some Beatles music. An enterprising deejay at WWDC, Carroll James, arranged for a BOAC (the major British air carrier) stewardess to bring over a copy of the English single, which had been released two weeks earlier. By December 17 Marsha was introducing the imported single on WWDC, and the ball was rolling.[3] The rapid dissemination of tape recordings to other radio stations pushed "I Want to Hold Your Hand" into the limelight, and although Capitol issued many cease and desist warnings to radio stations, it could not stop them from playing it. The record company was forced to move up the release date to December 26. From this point on the American fans managed to gain access to all Beatles records issued in England.

The development of more powerful and reliable vacuum tubes during the war had applications in many areas, but especially appropriate to the emergence of Beatlemania were the submerged repeaters produced by the British Post Office (GPO) in 1943. These extremely durable amplifiers dramatically improved transatlantic communication. Although the transatlantic telegraph cable was completed in the 1860s, telephone calls were routed through wireless radio networks when the service began in 1927. In 1956 a joint venture of the GPO and AT&T produced a transatlantic submarine telephone line that opened up telephone service between Great Britain and the eastern coast of North America. TAT 1 incorporated a lot of wartime technology, including submerged repeaters, coaxial cables, and new materials for insulation. It had only thirty-six channels, but TAT 3, which was laid down in 1963, had 138 channels, which made transatlantic calls much more accessible and cheaper. Brian Epstein made use of them arranging the Beatles tours, for, as Sid Bernstein said, they did everything on the telephone.

Government-sponsored research into sound reproduction was also applied to consumer products. The desperate fight to keep the Atlantic seaways open was the longest campaign of the war, and the underwater search for German submarines forced improvements in capturing their sound and reproducing it for the benefit of sailors who were learning to use location equipment like sonar. A team of English engineers led by Arthur Haddy developed a new way of making and duplicating master recordings called full frequency response recording, which captured more sound than existing methods. This innovation was incorporated into high-fidelity Decca recordings marketed after the war.

In EMI's research laboratories, the inventor Alan Blumlein continued his work on television signals and stereophonic sound, developing a portable radar set that would be a critical weapon in the war. He died testing the system when his RAF bomber crashed in 1942. Television also benefited from wartime innovation. Most of the development work was carried out in the research laboratories of the multinational entertainment companies, RCA-Victor in the United States and EMI/Marconi in England, during the 1930s. The postwar introduction of television owed much to the manufacturing techniques developed in wartime factories to mass produce vacuum and cathode ray tubes for radar sets. These innovations were applied to manufacturing cheap television sets, which were rolling off the assembly lines by the late 1940s, when governments began to allot wavebands and to license broadcasters.

No one exploited transportation and communication technology better than the Nazis. Hitler used air transportation to win elections and air power to cow his enemies. Propaganda minister Josef Goebbels called the new mass media of film and radio "the chief machinery of the intellectual leadership of the nation" and used it as a political weapon in creating a totalitarian state. After their initial victories in World War II, the Nazis had most of Western Europe to bombard with propaganda.[4] Weekly newsreels, public speeches, patriotic songs, educational messages, and radio newscasts were important weapons in Goebbels' war, and all of them could be recorded and rebroadcast.

Such was the burden of recording that the propaganda ministry commissioned research into a better way than the mechanical disc cutters that inscribed sound on delicate acetate discs. Giants of German industry AEG and I. G. Farben worked together on a system of electrical recording using thin tapes coated with magnetic powder. American soldiers brought captured Nazi tape recorders back to the United States and helped turn them into consumer products, manufactured by startup companies like Ampex and Magnecord. These companies also developed tape recording for images and introduced the first video recorders in the United States in the late 1950s and in Great Britain soon after.

The war effort brought about large-scale intercourse of ideas across the Atlantic and generated a lot of new knowledge. The Allies recruited scientists and engineers as well as sailors and airmen, and trained thousands of enlisted men in advanced technologies like radar and sonar. Others were inducted into the services to learn drafting or industrial management. These veterans went on to become technicians in recording studios, filmmakers, radio deejays, picture editors for newspapers, and engineers who devised better amplifiers. They also became the businessmen who set up independent record companies, musical instrument manufacturers, and retail outlets for electronic equipment.

Wartime experiences were especially influential in the film industry. Every combatant nation sent camera teams to the front to bring images of the conflict home, issuing 16 mm movie cameras to hundreds of enlisted men. These documentary filmmakers ended the war with the technical skills to move on; Gilbert Taylor, who shot film from a RAF Lancaster during nighttime bombing raids over Germany, was Richard Lester's director of photography on *A Hard Day's Night*.

The 16 mm cameras invented in the 1930s were redesigned in light of wartime usage and greatly improved. The Arriflex camera manufacturer introduced its 16 ST model in 1952, which became standard equipment for news gathering (and later for television news) and film studios. The new 16 mm cameras used smaller loads of film, which made them light and portable. Much faster film stock, like Kodak Tri-X, gave the cameraman more flexibility to shoot in low light situations. These innovations liberated the camera from the soundstages of film studios and permitted filmmakers to take them wherever there was action to be filmed. Wartime cameramen who worked in the film industry as civilians used fluid, highly mobile camera work to redefine the look of motion pictures, influencing French new wave cinema and the documentary style of the 1950s, both of which can be seen in Lester's film.

The development of a British amplifier industry was significantly affected by wartime experiences. The leader in the field was the American Fender company, whose products were extremely expensive but highly prized by English musicians. Dudley Craven was a young engineer in EMI's military research who went on to improve the circuitry of Fender amplifiers. He was hired by Jim Marshall, who had received technical training during the war and afterward ran a musical instrument store that imported American guitars and amplifiers. Marshall had taken apart Leo Fender's Bassman amplifiers and thought they could be duplicated without much trouble. With the help of his assistant, Ken Bran, Marshall and Craven produced the first Marshall amplifier.

Tom Jennings was another entrepreneur in the musical instrument business who thought he could make a better amplifier. He had befriended Dick Denney while they were working in the munitions industry during the war. When Denney brought a homemade amplifier into Jennings' shop, Jennings was so impressed that they formed a company to start manufacture. The first Vox AC 15 (for 15 watt output) amplifiers were introduced in 1957, and Vox followed them in 1959 with the AC 30, which became their most famous product and made Vox amplifiers synonymous with the British Invasion sound.

The development of amplifiers shows how quickly the technology could be improved within the constant interchange of British and American designs. Vox and Marshall started as followers, as copyists, but soon became innovators as technological leadership shifted to the UK in the late 1960s. They led the way in increasing the power output of amplifiers as rock'n'roll and larger audiences put more demands on them. In the beginning the Quarry Men used the small tube amplifiers found in radios and public address systems, for in those days you needed only relatively low output. Mo Foster's account of the early years of

British rock'n'roll is called *17 Watts* because this was enough for guitarists. Most car stereos these days produce more than 20 watts![5] Skiffle required little or no amplification, but this would not do for rock'n'roll. When the Beatles heard Eddie Cochran and Gene Vincent at the Liverpool Empire in March 1960, they were struck by how loud those Americans played. The Beatles knew they had to amplify their sound to play rock'n'roll, and they liked playing loudly. Critics who went to the Beatles' early concerts commented on the high volume, complaining that they were "loud beyond reason" and "too loud." With the assistance of Vox, the Beatles played louder and louder from 1961 to 1966, trying to overcome the screams that threatened to drown them out.[6]

Technological innovation not only increases the performance of machines, but can also be used to make them cheaper. When the Beatles entered their teenage years, guitars were not a common sight in the UK. Tommy Steele, looking back at the beginning of his career, commented: "I had a guitar—which was very rare at that time."[7] Most of these instruments were imported into the UK, chiefly by the Selmer Company, which had close ties to manufacturing facilities in Holland, Germany, and Eastern Europe, and that is why the Beatles first played guitars made all over the world, from Czechoslovakia to South Africa. But bolstered by the siren call of rock'n'roll, and supported by improvements in mass production, British manufacturers met the demand; guitar sales in England jumped from 5,000 in 1950 to 250,000 in 1957 and kept on rising.[8] The wave of amateur bands that formed in the wake of the Beatles was a result of the availability and affordability of electric guitars and amplifiers.

Wartime advances in sound technologies were translated into new products for the record industry, and there were so many amazing innovations that some historians have talked about a revolution in sound and credited the emergence of rock'n'roll to it.[9] The new 45 rpm singles and the 33 1/3 long-playing albums were smaller, more durable, and considerably lighter, which made them easier to transport. Arthur Kelly, a school friend of George Harrison, got hold of some 45 rpm vinyl discs that his brother-in-law had brought back from a business trip to New York, and he was astounded by the clarity of the reproduction. One of the discs was an EP of songs by someone named Elvis Presley. Before the first track was over, Kelly had called George on the telephone and asked him to come over.[10]

Many of the new sound technologies were the work of large corporate organizations in the United States and their research laboratories. The 45 rpm disc was introduced by RCA-Victor and the LP by Columbia, both companies who were part of the business and technological networks that stretched across the

Atlantic. They diffused technology through export, cross-licensing agreements, international financial investments, and the transatlantic movement of key personnel. They were drawn together in a web of mutually beneficial financial and technical relationships that made the diffusion of new technologies, like vinyl discs, and new entertainments, like rock'n'roll, relatively easy.

THE BEATLES AND THEIR MACHINES

If you look at any photographs taken of the Beatles on their tours, you will see them grasping many new high-tech products: the single-lens reflex cameras they hold as they descend from their airplane, the transistor radios stuck to their ears as they sit in their limousines, the record players in every hotel room, the portable tape recorders they carry in and out of Abbey Road studios, and the movie cameras strewn around their holiday beaches. While traveling by airplane across the United States, they listened to music on a portable reel-to-reel tape recorder. Wherever they went on tour, they picked up the latest sound and photographic equipment, for they were voracious consumers of new technology. In Japan representatives of electrical manufacturers visited the Beatles' hotel rooms to show them the latest products.

Their homes were full of high-tech kitchen equipment, the most advanced hi-fi systems, and numerous telephones. Paul McCartney's room in the London house of the Ashers (his girlfriend's parents) was described as being full of guitars, tape recorders, records, and phonographs. A visitor to John Lennon's mansion found five televisions turned on and record players in every room. Lennon described his first date with Yoko Ono as showing her his collection of tape recorders. John, Paul, and George built elaborate home recording studios when they moved into bigger houses in London.[11] Ringo liked hanging out at John Lennon's house "cos he's got all the toys there, tape recorders and things, which we like playing about with."[12] The Beatles certainly enjoyed their machines, at home, on the road, and in recording studios. They played with them and often talked about "having fun" with tape recorders and cameras.

All were enthusiastic photographers from their boyhood years, learning about photography with their parents' box cameras—a technological marvel of the older generation. Mark Haywood has an early picture of John Lennon in his large collection of Beatles photographs, a self-portrait taken when Lennon was about seven. Once rich and famous, the Beatles bought Pentax SLR cameras, 8 mm movie cameras, and projectors. The photographer Robert Freeman, who accompanied the Beatles on their visit to Paris in 1964, noted that they carried as much photographic equipment as the professionals who were taking their pictures.[13] The four musicians spent most of their professional lives surrounded by cameras, and the Beatles were one of the most photographed groups ever. When they were introduced to Tony Armstrong-Jones, husband of Princess Margaret and a professional photographer, at the Royal Variety Performance, he told them, "I must be the only photographer who hasn't photographed you."[14]

They were also exposed to documentary filmmaking from an early point in their careers. When they played the Cassanova Club in Liverpool's city center on Valentine's Day 1961, someone in the crowd used an 8 mm camera to make the first (silent) film of them. From then on they were always before movie and television cameras. They took the opportunity to peer down viewfinders, examine lenses, and check out light meters whenever they were in studios or on location. Their love of technology and their natural curiosity naturally drew them into filmmaking, and as early as September 1963 (before their film contract with UA was negotiated), both Paul and George had indicated their desire to get on the other side of the camera. By 1968, when the band visited the maharishi in India, they had to get special permission from the Indian government to bring in large amounts of 16 mm and still film they planned to use.[15]

Film was one of the Beatles' great shared interests, and cameras among their favorite toys. On their American tours they were surrounded by film cameras: the 16 mm professional models used by the news teams, who put their images all over television (there were no portable television cameras at that time), and by professional filmmakers like the Maysles brothers, who were making documentaries about the band. Then there were the 8 mm amateur cameras in the hands of the press, hotel staff, fans, and their entourage. The compact, inexpensive 8 mm movie camera was an example of those wonderful new consumer durables coming from America.

Like most of their peers in England, the Beatles had a positive view of new technology (especially if it came from the United States) and eagerly embraced the concept of modernity embodied in it. The Beatles were part of that generation of boys who had grown up on science fiction (like *Dan Dare, Pilot of the Future*), toy space rockets, and chemistry sets. They did not remember much about the technological horrors of the war (but their parents never forgot it), and they preferred postwar optimism without the nuclear anxiety that was ingrained in their contemporaries in the United States. "Atomic" had positive connotations for postwar Liverpudlians, as it expressed great power, such as Pete Best's "atomic drumming." The postwar world was described as the atomic age, the jet age, and the electronic age, but most of all it was an age of affluence.

The Beatles might have been very rich, but they were surrounded by the fortunate baby boom generation that could afford many of the new consumer products. As Prime Minister MacMillan made a point of telling them, they had "never had it so good," and mechanized entertainment was a big part of this new affluence—the radios, stereos, televisions, and cameras they were buying on credit and bringing home. The Beatles and their generation were part of a technological democracy that had unprecedented access to new machines, especially those that enhanced leisure time.

The most important element in this new technological democracy was the transistor, the electronic device that replaced the bulky and unreliable vacuum tubes in amplifiers—the heart of the machines that boosted sound and reproduced music.[16] Transistors were much smaller, did not heat up, and required much less current, so they could be powered by batteries. What the transistor wrought was a new generation of cheap, portable devices like cameras, radios, and tape recorders. These new products also benefited from manufacturer's increasing experience in molding brightly colored plastics, which made the devices more attractive to youthful buyers, and in mass producing electrical components, which brought down the cost. The first transistor radio, the Regency TR-1, was introduced in United States in 1954, but at $50 it did not sell well. Although the major electrical manufacturers like RCA and Philips were naturally in the lead (for they developed the technology), it did not take long for others to obtain the licenses to use transistors, and Japanese manufacturers such as Sony were experts at reducing the size and price of devices that employed them. The low price and portability of these machines made them perfectly suited for young people constantly on the move and requiring constant entertainment. They offered many of the benefits that make cellular phones a necessity of teenage life in the twenty-first century. And most important of all, these machines were operated by the children, not their parents. "Suddenly, we were in control of the radio . . . it was our own little thing that we carried around with the earphone."[17]

Social scientists have studied the behavior of teenage girls in the 1960s and described a "bedroom culture" revolving around the control of space in the American household.[18] Affluent Americans could give children their own room, a space for social interactions with other girls and boys: gossiping, reading teen magazines, and listening to music. These rooms contained transistorized radios and record players built for the 45 rpm disc. One of the great shared experiences of the rock'n'roll era on both sides of the Atlantic was listening to a transistor radio in bed when you were supposed to be asleep. In the United States, teenagers could also listen to transistorized radios in their cars—another private place for social interaction. The youthful audience became accustomed to listening to loud music through tiny speakers, which reproduced few lows but lots of sharp, tinny highs set against a humming background. Pop music record producers installed small radio speakers in their studios to discover exactly what the kids were hearing and mixed their recordings accordingly.

The bedroom culture of England was dominated by the Dansette, the portable record player introduced by Morris Margolin in 1950. The player came in a leatherette-covered case with a loudspeaker in the lid and a three-speed changer (33, 45, and 78 rpm), but it was predominantly used as a 45 rpm single player. Its compact size and bright colors appealed to girls, and it was so popular that Dansette became a generic term for small record players. In the United States similar machines mass marketed by retail outlets like Sears and Roebuck cost no more than $30. Loving the Beatles required a personal player: "All I wanted for Christmas was a record player . . . with the plastic casing and the little turntable." The holiday season was kind to Beatlemania, not just for the new Beatles album that made the perfect gift but for all the other machines that brought the band closer: "I begged for a little transistor radio that year."[19]

The transistor radio and camera were the machines that powered Beatlemania. The radios were not only in the hands of the Beatles but also in the clenched fists of the thousands of fans pursuing the group. "A lot of the girls had little transistor radios, so you could hear the music blaring," and you could also hear the deejays frantically maintaining excitement and telling everyone what the Beatles were doing and where they were going.[20] The radio stations set up remote broadcasting facilities at concerts or besieged hotels to communicate with the fans in real time. The transistor radio was an essential part of massing and directing the mob, but it also provided the technological links to give a shared sense of group identity with thousands of fans all listening together and interacting with each other through deejay intermediaries.

The transistor radio and Dansette were part of a new array of machines designed for youthful consumers: brightly colored radios, record players, hair driers, clocks, tape recorders, and cameras, all made from plastic. They were smaller and cheaper than the larger models sold to their parents. These machines were part of the youth culture of the 1960s as well as part of a broader twentieth-century movement of consumers capturing the bright points of their lives and recording them for posterity. This was a nineteenth-century idea, the basis to market Edison's talking machine in 1877, and soon there was a phonograph and George Eastman's box camera for every pocket. This reflexive saving of experiences reached a height in the 1960s, when transistors brought versatile recording technology to the masses. Beatlemania had the power to make the fans feel part of something important, and that is why they used the term "historic" to describe their experience. This made it essential to preserve it. Some of them recorded the sounds of *The Ed Sullivan Show* with tape recorders; others took pictures of the television screen. All wanted a memento of the live concert and took cameras with them.

While the Beatles were touring the UK in the early months of 1963, the American Kodak company was introducing a new product that would revolutionize amateur photography. The Kodak Brownie (introduced by Eastman in 1900) was instrumental in making photography a recreational activity for the masses, and its design made it easy for anyone to point and shoot. The Kodak Instamatic was built along the same lines, with fixed shutter speed and focus, and featured an innovative new way to take the challenge out of loading sensitive film into a camera. The Instamatic's 126 film was enclosed in a cartridge—the format that would make inserting tape into a recorder much easier—that could be dropped into the back of the camera. The first Instamatic 50 model was introduced in the United Kingdom in early 1963, and the 100 model followed in the United States. Priced at $15.95, the Instamatic was an instant success and went on to sell 50 million units from 1963 to 1970, and from then on all the small point and shoot cameras made by numerous manufacturers were called Instamatics. It was the ideal model for a beginner, and many baby boomers remember it fondly as their first camera. The American version, the Instamatic 100, had a holder for a flash-bulb that popped up from the camera body, enabling the user to take pictures inside; these were the cameras that fans took with them to concerts. When the Beatles appeared on stage, there would be an explosion of flash photography as well as an eruption of screams: "The flashbulbs were more impressive than the noise . . . the flash involved 55,000 thirteen year old girls all with these little instant cameras flashing like a strobe light."[21] Thousands of flashbulbs littered the auditoriums after the shows.

Instamatics, transistor radios, and binoculars were essential items to take to a Beatles concert. The equipment used by the press corps and traveling deejays were tape recorders and film cameras. Capitol's promotion campaign sent pre-recorded tapes of the Beatles answering a list of questions to radio stations all over the country., Deejays could pretend to ask the questions, and the tapes provided the answers, giving listeners the impression that these were personal interviews. Journalists and deejays used reel-to-reel recorders on the tours to preserve priceless interviews with the band and feed them into the broadcast networks by telephone. They accomplished this with the most primitive means, such as using alligator clips to connect recorders' loudspeakers to dismantled telephones, but this was the 1960s, and such feats of telecommunication were high tech. Each night on the tour, deejay Jerry Armstrong would send his Beatles' interviews on the telephone wire to WKYC, Cleveland, where the station edited the recordings and placed them in his radio show, promotional spots, and local news, before finally feeding them into the NBC radio network.

These technological systems spread the news of the band nationwide while giving priceless publicity to the agency that secured them. Murray the K followed the Beatles around, hollering, "What's happening, Baby? The Beatles are what's happening—tonight and every night on WINS" into his portable tape recorder. Larry Kane remembered lugging around heavy Grundig reel-to-reel recorders from performance to performance, then transferring the interviews into programming, and saving the best bits on tape cartridges to be used later. Kane even recorded conversations with the band on a portable recorder while they were in midair. The tape reels were sent to New York and then transported by air to England to be used in an awards presentation. The miracle of portability is revealed on the tape: "Hello people, this is John Lennon. Thanks for the award. We're flying over America you know . . . Paul speaking to you from four thousand miles away. Amazing what you can do with a small piece of tape."[22] The Beatles' management was well aware of all the things you could do with a piece of tape. They prohibited all tape recorders and film cameras from the American concerts, and with good reason: numerous bootleg films and recordings of their performances were in circulation. Pirates sold their unauthorized films of the Beatles to fans who already had 8 mm movie projectors.

Recording technology played a critical role in promoting the Beatles to the listening public. Epstein's strategy of media saturation would have been impossible without a means to save and rebroadcast their music, and the Beatles spent far more time taping radio and television programs in the early 1960s than they did recording songs to be made into records. Recording music instead of performing it live enabled Epstein to vastly increase the amount of Beatles product available to the media. As the hardest working band in show business, the Beatles were working full out from 1962 to 1966, and every minute of their output was recorded. Modern recording techniques enabled them to be here, there, and everywhere during Beatlemania.

SOUND RECORDING TECHNOLOGY

Tape recorders had a special role in the Beatles' creative process. The manufacturers improved them in the 1950s by standardizing tape width and speed and employing better materials for the tape. They applied mass-production techniques
to increase output, and subsequently tape recorders became more common in
English schools and homes. Cheap models for amateurs first appeared on the
market in the early 1950s, and one of the most popular and widely available was
the Grundig. The German radio retailer Max Grundig formed this company after
the war to manufacture kits for radios, and he soon moved into manufacturing
tape recorders, which he hoped to sell to amateur filmmakers who wanted sound
to go along with their 8 mm home movies.

Home taping had become a favorite pastime of technically oriented boys,
and by the early 1960s simple, inexpensive tape recorders were on the market
produced by Grundig, Telefunken, Philips, Ferguson, and Fidelity, which were
mostly large electrical manufacturers who saw tape recorders as another consumer product. Amateur recorders used their tape machines to save music or
just for the fun of recording something. Bob Molyneux brought his Grundig TK
8 reel-to-reel machine to the church fete at St Peters in Liverpool on that historic
day in July 1957. His was the first recording of the Beatles.[23] A tape recorder was
an important piece of equipment for amateur guitar bands. The Quarry Men
used a reel-to-reel recorder (removed from school by John Lennon) to save their
practice sessions. Allan Williams employed a Grundig tape recorder to promote
the Silver Beetles, but when he got to Hamburg he was not able to play the tape
of their songs because he had set the tape speed incorrectly.

The Beatles used tape recorders to help create their songs and gradually incorporated this expertise into their education as recording artists. We think of
them as song*writers*, but they could not read or write music. They wrote down the
chords and their sequence in the structure of the song, and they also wrote down
the lyrics, gradually accumulating enough pieces of paper to assemble a song.
Tony Bramwell noted that their songs began in exercise books and on scraps
of paper, which were always "floating around."[24] Remembering a melody or a
musical hook was the problem, but along with the professional songwriters of
Denmark Street, the Beatles believed that if you could not remember tunes after
one hearing, "they're no good anyway."[25]

As the Beatles grew in confidence, and their songs became a little more complex, they needed something better than bits of paper to save their work, which
led them to sound recording. They began to assemble their songs out of guitar

phrases that were recorded rather than written down, and they took the recording media with these fragments (on tape or disc) into their homes and recording
studios as a memory aid in the creative process. John and Paul rarely composed
music in the classical manner, by sitting down at a piano, but instead collected
and arranged musical fragments. They gradually built up songs by configuring
parts like middle eights and working out the harmonies. It was rare that a Beatles' song emerged from either Lennon or McCartney fully formed; instead they
would develop it over time with numerous additions and subtractions as the two
contributed whatever they felt appropriate. Sometimes it might mean changing
a line; other times it might involve rearranging the melody. It was a collaborative
process. Tape recording allowed Lennon and McCartney to remember all the past
combinations while generating new ones. Even the process of tape recording—
recording, playback, replay, recording—fit into their work habits and gave them
time to evaluate their progress as the song evolved in the recording studio. They
fabricated recordings rather than wrote songs.

As amateur musicians the Beatles started recording at home and then took
this experience into recording studios. As their proficiency and competence in
studio recording increased, they found more ways to incorporate their home tape
recorders (and later home studios) into the creation of their music. They could,
of course, use the existing acetate technology to cut a disc, but this was time consuming and tricky, and as they worked longer hours at Abbey Road, the means to
do so were often not available to them. Their own tape recorders were the answer.
They saved musical ideas at any time of day or night, and these could be taken on
tape to the studio the next morning. The flexibility of tape recording gave musicians the ability to shift the location of the creative process, from the company's
studio to the hotels, tour buses, and houses where they usually composed their
songs. Over time several of the Beatles built their own recording studios, bringing together tape recorders and mixers so they could go beyond merely making
copies and instead carry out all the work they did at Abbey Road. John told an
interviewer, "I've bought all the taping and mixing equipment so I can add things
and do a lot more."[26]

Mastering the recording process was critical in gaining control of the means
of production in the music business, which in the early 1960s was almost exclusively in the hands of the major companies. The technology of making recordings with disc cutters was expensive, difficult to operate, and under the control
the company's employees rather than musicians. The acetate demo disc was the
standard in the transatlantic music industry to present musicians or to get a song
published or recorded. Record stores and artist management set up facilities to

cut discs to be used as demos, and some independent record companies had their humble beginnings in these small operations. Numerous commercial disc recorders were also in booths where you could record a song and take it away on a thin plastic disc. When Ringo talked about his ambition to make a record, he meant a record "that you hadn't made in a booth somewhere!"[27] The Quarry Men's first record was a version of "That'll Be the Day" recorded on an acetate in the studio of Percy Phillips in Liverpool, whose "professional tape and disc recording service" (advertised as the only one in Liverpool) was in one small room in his house. Philips had a Vortexian tape recorder, a MSS disc cutter, and some microphones. There was no dubbing or re-recording.[28] While the Beatles were in Hamburg, they used a small studio to cut an acetate of "Summertime," which was transferred onto a 78 rpm disc. Making acetates was expensive and required hiring professional technicians; tape recording, on the other hand, could be done by the musicians themselves.

As they became more experienced with home recording and more affluent, the Beatles acquired more advanced tape recorders than the Grundigs and Fidelitys they had started with. Small, technologically driven concerns like Ferrograph, Truvox, and Brenell were formed on the foundation of wartime technology and expertise. Two émigré Czech engineers who had served in the RAF established the Brenell company in 1947. They designed their recorders for the home user with considerable technical skills because their machines were sold as components that had to be assembled with the help of circuit diagrams. As home taping became more popular, Brenell introduced fully assembled recorders with the added benefits of variable tape speed and "magic eye" sound level indicators. Instead of requiring a connection to a valve amplifier, these tape decks had pre-amps built in. In 1958 they introduced the Brenell Mk.5, which became their best-selling model. They continually improved it to bring it up to professional standards, with features such as voltmeter peak recording-level meters, mixing facilities to bring together inputs from microphones, and full frequency equalization. In 1961 they brought out the Mk.5M, which came with three recording/erase heads and a superimpose function, which lifted the tape from the erase head and permitted re-recording over existing tracks—the critical function that made this machine a favorite with recording engineers and musicians. Brenell introduced its STB series in 1963. These advanced recorders had all the features found in a professional recording machine, and the STB2 of 1964 was a mini-recording studio.

Lennon and McCartney were coming of age as recording artists when these machines were made available, and they quickly acquired them. These "trusty

Brenells" were constantly in use as the musicians came and went from the studio, producing the shards of songs and musical ideas that were the raw material of their creative work. John and Paul talked about making singles as creating three-minute stories like the ads on TV, just that their stories were fabricated out of pieces of tape. When it came to the more ambitious recording projects like *Sgt Pepper*, the Beatles leaned more heavily on their personal tape recorders to generate the parts of the song that were later brushed up and joined in the studio. The Abbey Road engineers were often called upon to marry these pieces without altering the pitch and tempo of the final recording, and their efforts were critical in producing such acclaimed and complex recordings like "A Day in the Life." The memorable moments of genius might have occurred in the Abbey Road studios, but the creative process often began with a musical idea saved on a home recorder.

CHAPTER 11

IN THE RECORDING STUDIO

The Beatles' work in the recording studio from 1962 to 1969 is the best documented creative journey of all time. They started out by covering some ancient pop songs, such as "Besame Mucho," and went on to create some of the most acclaimed and remembered music of the twentieth century. When they first went into a studio in 1962, pop music was recorded on two-track machines, and each session was meant to produce two sides (of a record) in three hours. Recording was done as quickly as possible and under the control of the management. By the time the Beatles disbanded, eight-track recording was the norm, sixteen-track technology was on the horizon, and an album might take months or even years to produce; such was the amount of manipulation and re-recording that went into producing a single song.

The Beatles went into their first recording sessions ready to record whatever songs had been chosen for them, but by the time they made *Revolver,* they were using recording technology to create the music in the studio. Lennon and McCartney have been universally recognized as great songwriters, but their reputation as creative artists emerged from their mastery of all the technological tools of recording. The critic Paul Saltzman called them "the first poets of technological culture," and their management underlined the point that "the Beatles are still the creators" in the recording studio. Their fans remembered them as the great innovators who "revolutionized recording technology, instrumentation, fashion, hairstyles."[1] The Beatles turned pop music into art, and they did it in a recording studio.

John Lennon and Paul McCartney's impressive output of beautiful music did not rest on any knowledge of recording technology, because this was limited. They were not technically literate; in fact the engineers they worked with remembered them as "technologically challenged." A journalist who visited John Lennon in his home was impressed by the number of high-tech machines there, but Lennon admitted that only his gorilla suit worked.[2] The people who actually operated the controls and wired up the devices in recording studios knew that the band did not understand how the equipment worked.

The Beatles' lack of technical acumen was demonstrated in their dealings with "Magic Alex," a smooth-talking entrepreneur whom they commissioned to design a state-of-the-art recording studio in the new Apple corporate offices. Magic Alex promised them a seventy-two-track studio, but all they got for their money was a pile of junk that the Abbey Road engineers ripped out and replaced with borrowed equipment from EMI. The Beatles do not seem to have practiced much British craftsmanship in modifying or maintaining their guitars and amplifiers, which appear in pictures with broken parts, screws loose, and sticky tape holding vital components. Ringo had a habit of leaving his pack of cigarettes on the drum heads, which is not the way to treat drums, although it did produce some interesting sounds.

The success of the Beatles as recording artists came more from their intellectual curiosity and sense of playfulness than their mastery of the technology. Their reputation as purveyors of advanced technology and their success in applying it to create new and beautiful sounds rests more on attitude than on understanding the machines they used. They were lucky to be in recording studios when the technology was advancing rapidly and more and more music could be crammed onto a one-inch-wide recording tape. As creative tinkerers, they were often imaginative, bringing equipment and methods together in innovative ways, and they pushed the boundaries of what was technically possible. At the same time they changed the culture of the recording studio, and this innovation was almost as significant as the music they produced in this controlled space.

The Beatles were overawed by the large corporate studios of Decca and EMI. Just being there was an honor and a mark of their progress as professional musicians, yet the experience must have reminded them of their lowly status as small cogs in a large and rigidly organized machine. We always associate the Beatles with EMI's Abbey Road Studios, but in 1963, they probably spent more time in the BBC's Paris Studio off Regent Street in Central London than they did in EMI's North London facility. The Beatles were obliged to make five recordings every week for the BBC, which they accomplished in a busy three- or four-hour session at the Paris Studio.

The BBC did not have complicated recording equipment, only single-track mono fed by a few microphones, and there was no time to re-record parts that were not satisfactory. So the band had to get it right the first time, and it probably helped that they often recorded in front of a live audience brought into the Paris Studio, which had been built as a cinema, with plenty of seating and a raised stage. The other BBC studio used by the Beatles, the Playhouse Theatre, was also an impressive auditorium. Built in the 1880s with ornate decoration and regal boxes at the sides of the stage, the Playhouse evoked the golden age of the music hall, but now it was the originating point of a modern entertainment that would be broadcast worldwide.

By later standards, the recording technology employed by the Beatles when they first entered Abbey Road was primitive. Unlike the Paris Studio one-track machines, in which all the sounds are conveyed in one channel to the recorder, Abbey Road had two-track recording, which used two channels to produce either basic stereo or two-channel mono. Bands like the Beatles, who had only worked in mono, found that the addition of another track opened up all sorts of wonderful opportunities for recording over previous material (called overdubbing), and for adding strings or more vocals to an existing song. Recording two similar tracks of a vocal together (called double tracking) produced a thicker and more choruslike vocal. This was a popular sound in American recordings, as were the echo and reverb, which were also added as the track was recorded. The echoing lead vocal on records like "Heartbreak Hotel" made a big impression on the Beatles as it did on everybody else who listened to it. It was a signature sound, what Peter Doyle calls "a defining production characteristic of rock'n'roll music" in his book on echo and reverb.[3] Echo was used extensively in American studios, but the Beatles did not adopt it because it was too complicated and expensive. There was debate within the band whether to incorporate it and some relief when they realized later that not using it was the right decision because it made them sound different. The lack of echo became a mark of originality.[4] In his analysis of the Beatles' sound, historian of rock'n'roll Charlie Gillett points to the lack of echo and double tracking, plus the simple arrangements of their records, which gave a sense of authenticity that endeared them to the English fans. He contrasts this to the "lush contrivances" of overproduced American records.[5]

The Beatles stayed well within the technological boundaries set by the company for the entirety of their career with EMI, and the equipment they used was never at the leading edge of recording technology. EMI's Abbey Road facility is today the most famous studio in the world, a brand name in its own right and a major tourist attraction. But during the 1960s it was known merely as EMI's studio, and nobody who worked there thought of it as particularly advanced. Neither George Martin nor Geoff Emerick will argue that any British studio was technically superior to those used by the American majors. When Martin went to Los Angeles to observe Capitol's engineers recording Frank Sinatra's *Come Fly With Me*, he noted gloomily that the Americans had the best equipment and the most advanced techniques.[6] The Americans were clearly in the lead. Abbey Road staff, like all their colleagues in the British recording industry, examined imported records to figure out how to duplicate the new sounds coming across the Atlantic. Capitol's studio in Los Angeles—an architectural landmark shaped like a pile of records—had eight-track tape-recording machines, while in England the norm was only two or three track.

The mutual envy between English and American studios was the driving force in the development of the Beatles' music, especially in their creative competition with Brian Wilson and the Beach Boys—Capitol's other star group, who were recording complex and innovative albums in Los Angeles. An important moment in the history of the Beatles was when Paul brought a copy of Brian Wilson's seminal *Pet Sounds* into Abbey Road and asked the staff if they could duplicate that "clean American sound."[7] English musicians and technicians were also impressed with the loud and distinctive bass and drum sounds coming from Tamla-Motown records from Detroit. These too were copied in the production of the Beatles' sound.

When the Beatles started their musical career, pop music was considered so unimportant at EMI that it only warranted two-track machines—the four-track recorders were reserved for the classical acts and more serious music. It was a breakthrough when the Beatles were allowed to record on four tracks, and they only graduated to eight-track recording at the end of their careers. EMI was slow to adopt eight-track recording, and around 1968 the Beatles started to use other, more advanced studios in London, like Olympic and Trident, which had more tracks to record and better noise-reduction technology.[8] Nowadays a home computer can easily handle twenty-four recording tracks, but even in the high-tech studios of the 1960s, this would have been an impossible dream.

George Martin and his subordinates at Abbey Road made this relatively primitive level of two-track recording work well for the Beatles. They could alter the frequency levels of the two channels of sound by adjusting the levels on the mixing board in the studio's control room. (Each channel of sound is divided into frequencies—high to low—and the controls on the mixing board adjust the amplitude of each frequency as it goes into the mix of tracks on the record. This process is called mixing, and the final assembly of recorded sound is called the mix.) They could use filters or compressors to emphasize a specific frequency (a crisp high or a booming low) in the mix. They could use the two tracks to overdub pieces of sound onto an imperfect track, double track a vocal, or add new sounds. For example, the rhythm track of "Love Me Do" took fifteen takes to complete, and the final version was edited together from these tracks. You could add more tracks to the mix by "bouncing," or squeezing, tracks together, opening up a track for something else, but this was time consuming, and making recordings of re- cordings inevitably meant a loss of fidelity and increased tape hum.

The mammoth February 11, 1963, studio session, in which they recorded an entire album of songs in a day, marked the end of an era. This was the old way of making records, throwing it down live in one take using a couple of micro- phones and leaving the engineers to clean up the mixes by overdubbing flubbed parts. The introduction of four-track machines in October 1963 definitely made the work of recording the Beatles easier and gradually influenced the sound of their music. Having more tracks to work with meant it was possible to add more instruments, and different sounds, to the mix. So instead of one track for vo- cals and one for rhythm and lead guitars, a producer could use a track each for vocals, guitars, and drums, with one track left for a piano, a harpsichord, or even a string quartet. Multitrack recording made adding additional sounds easier and produced cleaner results. In the year after the Beatles started to work with four-track recording, we can hear more sounds added to their records, including exotic instruments such as the Indian sitar. Their songs were put together take by take over long periods of recording activity, in which the best of all the many different recordings were joined in more sophisticated and invisible ways. This process gradually increased the number of studio hours it took to produce an acceptable mix.

CHANGING STUDIO CULTURE

Working in a recording studio could be a daunting experience for young musi- cians who had recently given up their day jobs. There was something unnerving in the way the engineers listened impassively to the sound that came out drier and clearer from studio monitors than it did from the concert stage. As nervous musicians listened to their playback for the first time, it seemed so different (and amateur) compared with the sound that came back to them at a perfor- mance. The Beatles' confidence and relaxed demeanor were great assets because this outlook helped them overcome the nervousness that often paralyzed semi- professional musicians in the studio. They did not take it too seriously, laughing and joking and generally having more fun than the other rock and pop acts. Geoff Emerick thought they were much more relaxed in the studio than Cliff and the Shadows and remembered them as constantly messing around and be- ing silly. Another engineer who worked with them, Glyn Johns, recalled them being hysterically funny; he could not stop laughing as he recorded them. A journalist from *Mersey Beat* also noted their carefree attitude in the studio, as they laughed at each other's mistakes and joked around.[9] As BBC radio producer Terry Henebery remembered, "They'd come to the studio and horse about. You had to crack the whip and get on the loudspeaker talk-back key quite a lot and say 'Come on, Chaps!' They'd be lying on the floor, giggling."[10] But after locking each other in the toilets and fooling about for hours, the band always pulled everything together and produced good takes before their time ran out. They were professionals, but their playful attitude toward studio work gave them an advantage in creating new sounds.

Another challenge to the status quo came in the selection of material they were to record—a vital part of the process that had always been in the hands of the company A&R men. Once a contract had been signed, the company started to look for suitable material, and George Martin had chosen a bland but catchy pop song called "How Do You Do It" that he was sure would go to Number 1. He was right. Gerry Marsden took it to the top of the charts, but the Beatles were reluctant to put out something so lightweight. In the to and fro in the studio, the band agreed to record it but persuaded Martin not to release it. Martin told them, "When you can write a song as good as that one, then I'll record it," and the Beatles took him at his word and gave him "Please Please Me." Their assertiveness took everybody by surprise: "They've got some cheek, that lot!" said Norman Smith in the control room.[11]

The recording studio was run by the record company. Employees of the company selected the songs to be recorded, booked the musicians, approved the arrangements, and decided what takes would be used. We know that the Beatles had creative input in their recordings, but we forget that few bands had this privilege in the early 1960s. The Big Three was another Liverpool band managed by Brian Epstein and under contract to Decca. One band member sadly recalled, "We were never allowed to choose our own material." The company released a demo tape of the band as a single without ever consulting them because it was considered good enough to make into a master. This was the norm, not the exception. Another Liverpool group was appalled to hear that the company had added strings to their first record, which they had thought would be released as a country and western song.[12]

The company's technical staff placed the microphones around the studio, operated the recorders and mixing consoles, and supervised the recording. The post-production work of mixing, mastering, and disc cutting (in which the final version is made into a master record) was carried out by company employees, who also decided the running order of songs and the cover copy and art for LPs.

Musicians had a limited part to play in this process and were subject to the strict rules of demarcation between who did what in the recording studio. This was especially important in Great Britain, where unions had great power, and it only took a worker doing a job intended for another to provoke a strike. (A member of the Abbey Road staff who touched a cable in Twickenham film studios almost brought production of a Beatles' film to an end.)

To distinguish their roles, employees of Abbey Road wore uniforms that defined their jobs: white coats for the engineers, brown coats for the maintenance crew, and shirts and ties for upper management and slightly more prestigious jobs in the control room, like operating the mixing consoles. Balance engineers in suits would move the faders on the consoles but would not dirty their hands setting up microphones; instead they told the white coats where to place them. Brown coats could make minor adjustments as they connected cables on the studio floor, but only white coats were allowed to alter signal routing (from microphone to mixing console) or change the cabling. Musicians could not touch any equipment in the studio and were expected to do exactly what they were told.

EMI had rules for everything, both written and unwritten, as well as strict etiquette in recording. For example, lowly tape operators (called button pushers) were not permitted to alter anything on their own initiative, nor were they allowed into certain spaces, such as mastering rooms. Balance engineers (the next step up from tape operators) would not dare offer an unsolicited opinion to a record producer or an A&R man as they sat in the control room listening to the tracks being recorded.[13] There was a dress code and an institutional atmosphere where everyone knew their place and no one dared to challenge the status quo. Every Monday morning, the handsome wooden floors of Abbey Road were waxed with military precision. One famous visitor to the studios saw so many employees in white coats that he thought it might be a hospital.

The most important rule of the recording studio was that musicians were not allowed to interfere in the recording process. The first studio was established by Thomas Edison in his West Orange laboratory in the 1880s. He put up a curtain between artists and the acoustic recording horn of the phonograph, but as recording technology became increasingly sensitive to sound, more elaborate and permanent divisions were erected between musicians and the recording machine. As the technology developed in the 1930s and 1940s, studio engineers installed the recording and mixing devices in a control room, an insulated booth normally overlooking the studio space, with a large glass window through which eye contact could be maintained with the musicians. Artists were not allowed to enter the control room or trespass on technical activities. Abbey Road's Studio Two had twenty-two wooden steps between the floor of the studio and the control room. So on the day the Beatles were invited up there to listen to a playback, there was a sense of breaking with tradition. Paul's response was "control room, what us? Up those stairs, in heaven?"[14]

As befitting any industrial workplace, there were also rigid constraints on time, which was one of the most important factors of production in commercial recording. In the early days a stop watch was the only measuring equipment in the studio—because of the limitations of the recording media, it was imperative to know how much time was left on the master cylinder or disc. When the first rock'n'rollers strolled into studios, the basic unit of studio time was the three-hour session that was expected to yield two A sides, and often a B as well. A large facility like Abbey Road was booked throughout the day, and different musicians would appear at each time slot and be allotted one of the studios according to the size of the ensemble. The major record companies designed their recording studios to control the creative process and to mass produce recordings.

The Beatles entered Abbey Road in shirts and ties and dutifully did what they were told. George Martin arranged the record around their ideas, changing parts around, deciding on the tempo or the introduction, and enforcing his suggestions. It was his job to bring out the hooks and other distinctive features of the Beatles' sound to give them a musical identity in the crowded pop record market. To this end, Martin gave John's harmonica prominence in the first Beatles releases; its distinctive sound "almost jumped out at you" from the mix.[15] Martin gently pushed the band into faster tempos and more vigorous playing, and he was responsible for the engaging sound of the early singles, mixing the tracks to give them the characteristic bright, trebly register. From the dramatic ringing chord that starts "Eight Days a Week" to the triumphant, ascending introduction to "I Want to Hold Your Hand," the Beatles' songs always seemed to come at you louder and faster than the competition. Martin had enough experience in the pop music business to recognize the importance of a strong start. As he suggested ways to rearrange "Can't Buy Me Love," he told them, "We've got to have an introduction, something that catches the ear immediately, a hook."[16] George Martin's skills as a record producer (and as a musician) were as important a part of the Beatles' sound as their guitars and amplifiers. In songs like "Please Please Me," he played a much larger role than director or adviser; he made Lennon and McCartney restructure the song and increase the tempo to change a slow, bluesy number into a jaunty, danceable hit. At the end of the session, the voice in the control room asserted its control of the process when Martin told the musicians that they had made their first Number 1 hit record.

As the band learned the ropes of recording and became accustomed to the process of mixing and editing takes, they began to feel more comfortable and think of the studio as a workplace for musicians as well as for the men in white coats. The staggering sales numbers they were compiling increased their status beyond that of boy band. They were feted by Sir Joseph Lockwood and the rest of the company's upper management because the revenue from their records drove up profits by 80 percent during Beatlemania. As George Harrison pointed out, "We gained more control each time we got a Number One . . . we'd claw our way up until we took over the store."[17] A million advance orders for "I Want to Hold Your Hand" impressed even the most conservative company executives. They gave the musicians more say in deciding what songs and which takes to use and allowed the players to gradually exert more influence on the decisions about mixing them, which were normally made in the control room. The Beatles were still ingenues in the studio, and when Martin scoffed at the corny sixth chord that ended "She Loves You" on the last "yeah," they could not see his point. Their response was "It's a great chord!" and so it stayed in. If Joe Meek or Norrie Paramor had been in the control room, that sixth (and any connection with the big band era) would have been removed without any discussion.

A combination of the Beatles' spectacular sales figures and their scouser indifference to authority enabled them to transform the culture of EMI's studio. Things were indeed changing in the company. In an unprecedented move, EMI decided to save all the session takes of the Beatles, including the outtakes. This included not only the incomplete takes but all the mistakes, chatter, jokes, and silly noises the band produced. As George Martin commented, it was hard enough to archive master recordings at this time, so deciding to save everything the Beatles did in the recording studio was an amazing step. But there was method in this madness; the management had realized they were sitting on a gold mine, and every little bit of the Beatles on tape would have commercial value in the decades to come.

TURNING A WORKPLACE INTO A LABORATORY

The introduction of four-track machines influenced the Beatles' recording career in subtle ways. Records were still made by editing together the best takes and joining them into a final version, but more tracks allowed the technicians to make changes quickly and easily. The musicians found that moving from two to four tracks in the mix opened up their musical horizons. The engineer Ken Townshend concluded that the introduction of four tracks "made the studios much more of a workshop," and it led the Beatles to think of the studio as a place to rehearse and experiment.[18] The transition from workplace to workshop changed the Beatles' creative habits. Their recording of "Eight Days a Week" in October 1964 marked the first time they took an unfinished article into the studio and experimented with it until they got the final version they wanted. The session lasted from 3 to 10 p.m. McCartney brought in the song, and Lennon worked with him to fill in the middle eight and come up with an introduction. The worked through five takes and considered the sixth to be very promising. Seven takes later, they had take six perfected and the song almost complete. They recorded the distinctive faded-in introduction and a new ending several days later. The band was taking longer and longer to produce a take deemed worthy of saving. It took over eight hours to get two takes of "We Can Work It Out" in the can.[19] Now they were putting together a song over several days rather than finishing it in one session. This was an important change that gave them time to think about a piece and come back to it later with fresh ideas. In the opinion of Geoff Emerick, this gave them room to be more creative.

It was a short step to turn a workshop into a laboratory. The Beatles began to experiment with the electronic effects that intrigued all musicians who played heavily amplified music. The amplifier feedback that distinguishes the beginning of "I Feel Fine" has been described as an "electronic accident" and was the first time feedback was used on a pop record.[20] John Lennon was not alone in using feedback creatively, and "I Feel Fine" does not mark any great innovation in guitar sounds, but the experiments that led to the finished record were significant because they reveal the changing attitude of the musicians. Employing amp feedback, single-echo delay, and distortion from the "fuzz boxes" built by Abbey Road staff shows that the Beatles felt comfortable enough to experiment with the tools at hand and powerful enough to commission new ones to play with. During 1964 the relationship between the Beatles and their recording manager was changing. George Martin was no longer dictating the proceedings. He had learned not to dismiss any of their ideas out of hand—too many of them had been winners—and was prepared to give them more freedom in the studio to experiment.

The development of the Beatles' music and the corresponding freedom to challenge the status quo at Abbey Road is evident in 1965. This was the year they recorded *Rubber Soul*. Songs like "Norwegian Wood" and "In My Life" marked a departure from the music that had made them superstars. More personal, introspective, and complex than "She Loves You," these recordings revealed a new maturity. The Beatles were now taking control of Studio Two at Abbey Road, inviting their friends in, drinking and smoking there, and staying well past midnight. In his detailed analysis of the Beatles' recording career, Mark Lewisohn viewed 1965 as a time of important innovations in the musicians' studio work habits. They were now using tape recording in new ways to improve their techniques of writing and recording a song. They were rehearsing songs with the tape running and then spooling back to record over the initial run-through. By adding numerous overdubs onto a basic rhythm track, they reduced the number of complete takes required of them, easing their work but increasing the load on the engineering staff who had to dub in the many additions.[21]

The recording of "Yesterday" in the summer of 1965 marked another first in making pop records. The Beatles had previously introduced instruments like timpani and flute into their sound, but "Yesterday" involved a string quartet (two violins, a cello, and viola), which gave the record more than just a hint of classical music. The technological underpinning for this development was the ability of multitrack recording to accommodate more channels of sound and thus more different types of sound, but beyond this were more changes in the culture of the recording studio. At Abbey Road, as at other corporate studio complexes, there were cultural boundaries between the different genres of music. EMI built Abbey Road to record large orchestras, and the musicians and the technicians who worked with them were considered quite snobbish by the people who worked with smaller, uncouth pop groups. From "Yesterday" on the Beatles employed more classical instruments in making their records, and Lennon and McCartney had more interaction with professionals whose day jobs were in symphony orchestras. The same process of give and take went on in the studio, but now it was with classical session men rather than with engineers or office managers who had to accommodate the Beatles. The Beatles' ignorance of what could and could not be achieved by an instrument worked in their favor, for their wild ideas sometimes panned out. Paul did not always get the sound he wanted out of string quartets or French horn players because he often asked for the impossible, but the musical compromises between composer and musicians led to a creative encounter between the new and the old.

A similar transition occurred in the control rooms of Abbey Road, where younger technicians like Geoff Emerick were now at the consoles and chaffing at the rules. They were open to experimentation, and as a tape operator concluded, "because the Beatles were very creative and very adventurous, they [the engineers] would say yes to anything." When Emerick took over as their engineer, "experimentation became the key," and he called the chapter of his memoirs that recounts the making of *Revolver* "Innovation and Invention."[22] The Beatles' brazen disregard of EMI's regulations emboldened the younger staff at Abbey Road so that they began to challenge the status quo. They broke the rules about how close you could place a microphone next to an instrument, especially drums. They challenged the construction of space in the studio, entering the province of other engineers and crossing the rigid boundaries of demarcation. Just like the Beatles, who by 1966 were getting involved in the mixing and editing of their master tapes, the technicians were also reaching beyond their place and trying to influence processes, like mastering, that were off limits to them. These cracks in the cultural organization of the studio paid dividends because it led to more creativity. Automatic double tracking (ADT) was one of these valuable innovations. The engineer Ken Townshend invented this process because John Lennon was a one- or two-take man who did not want to continually double track his voice. ADT solved the problem and became an important component of the Beatles sound.

The *Rubber Soul* LP marked the decisive step toward the pop music album as an entity, a carefully constructed artistic whole rather than the collection of singles it had always been. *Revolver* continued this trend. By the time of *Sgt Pepper's Lonely Hearts Club Band*, experimentation was the norm in the studio, and creativity was at a peak. The Beatles were masters of imagining new sounds, but they did not have enough musical or technical expertise to realize them; instead they relied on the staff of Abbey Road studios (George Martin, Geoff Emerick, Norman Smith, and Ken Townshend) to turn the sounds in their minds (often numb with drug use) into something that could be recorded. Thus, when John Lennon told them, "I want my voice to sound like the Dalai Lama chanting from a mountain top, miles away" (for "Tomorrow Never Knows"), they had to use their initiative to come up with an answer. Geoff Emerick got the idea of taking out a Leslie speaker from a Hammond organ and inserting it into the signal chain. Ken Townshend did the wiring.[23] The distinct sound of Lennon's voice and Ringo Starr's drums owed a lot to the technicians, and their experiments in double tracking Lennon's voice or close-miking Ringo's drums are now standard procedure in recording studios. The Beatles did not discover or understand the process; all they invented were the names—so that ADT is now called "flanging," after Lennon's nickname for it. I am not saying that the technicians could have made the music without the Beatles, but that the Beatles could not have made the same music without the innovations of the technicians.

George Martin was now the facilitator of their creations rather than the director he had been at the time of "Please Please Me." The important advances in making the bass guitar more prominent on the record, in cutting masters, in developing vari-speed techniques (in which the spinning capstans of tape recorders were slowed down or speeded up)—all came from requests or ideas from the musicians themselves. Now they were at the controls of the mixing consoles, busily editing their own music (at least they were playing with the faders on the mixing consoles—Geoff Emerick is convinced they made few changes), and interfering in processes—like mastering—that not even the mixing engineers were allowed to do. The statuses of musicians and company men were now inverted, as George Martin recognized: "We were building sound pictures and my role had changed—it was how to interpret those pictures and work out how best to get them down on tape."[24]

The critical tool in their experiments was magnetic recording tape. All that playing around with tape recorders yielded some important results. Once the Beatles familiarized themselves with operation of their trusty Brenell recorders, they started to experiment with tape loops—lengths of tape joined together—and this brought them to some music that was so completely different from what they created in the past that John and Paul could think of themselves as avant-garde. "Revolution No. 9" was a step into the brave new world of electronic and sampled sounds and a forerunner of today's loop-driven dance music.[25] "Tomorrow Never Knows" was built around five tape loops running simultaneously to recreate the experience of LSD. This song contained a variety of tape-based effects and was a radical departure from anything the Beatles or any other pop band had done before. As the last track on *Revolver*, it signaled a new phase of experimentation and innovation from the Beatles. From its inception, Lennon reading lines from Timothy Leary's *The Psychedelic Experience* into a tape recorder, to Paul's guitar solo recorded backward, tape-recording technology was integral to the record's fabrication. The Beatles continued their love affair with tape recorders, being among the first to use the new tape cassette players and always finding new ways of using pieces of recording tape to make music.

Of all the innovations credited to the Beatles in the recording studio, the most important was their example to other musicians. The rise of singer-songwriters along with the Beatles' mastery of multitrack recording not only changed the sound of pop music but also elevated the status of its creators. The Beatles made fellow musicians aware of the potential of recording technology and its subordination to their artistic desires. Their highly produced records were examples to be followed by many others, and their success in selling them brought more entrepreneurs into the recording business, loosening the grip the two major record companies had on commercial sound recording.

EMI treated George Martin as just another employee and refused to give him a share of the immense profits he generated for the company, so he left to form his own AIR Studios with some top producers and engineers he took with him from Abbey Road. More independent studios were formed in the late 1960s in the wake of the Beatles. Some were new entries in the field, like Trident studios in Soho, and others were rebuilt and reorganized around small existing studios, such as Olympic Sound Studios in Barnes. Some had the newest equipment; others were small and dirty but had character, like Regent Sound Studio on Denmark Street.[26] The Beatles helped end the domination of the big corporate studios built by EMI and Decca. George Martin now worked for EMI as an independent producer, and the Beatles recorded wherever they pleased. The musicians who a few years ago would not have dreamed of walking up to the control room were now building a recording studio in their corporate headquarters.

The Beatles were now in complete control of the Abbey Road workplace, and the technical staff was at their beck and call. Musicians were free to come and go as they pleased, free to take drugs in the studio, and free to break the unwritten law and bring their girlfriends into this masculine space. Things were so out of hand by 1969 that John even went so far as to move Yoko Ono's bed into Studio Two while she was unwell—perhaps no single act could have been calculated to do more damage to the unity of the band. By 1968 those engineers who had been closest to the Beatles were beginning to leave. As the antagonism within the band increased, more staff refused to work with them, turning down an opportunity that would have been unthinkable to do in earlier years. There was a tense and unpleasant atmosphere between band and technical staff, an "us versus them" situation. The musicians let everyone know how much they hated working at Abbey Road, and as Geoff Emerick points out, that famous image of them crossing the road outside the studio has them walking *away* from it.[27] The secret hidden in this image was not that Paul was dead, but that the Beatles no longer wanted to record together. They had made the record the raison d'être of the band. As John Lennon said, "We'll still be pop stars as long as we continue making records."[28] Once this function had been lost, there was no longer any point in being Beatles.

CHAPTER 12

THE BEATLES AND THE SIXTIES

Even after the band broke up, the Beatles' media presence remained powerful. They were still in the public eye in the 1970s; a young fan recalled, "As an American they came into your life automatically without you lifting a finger."[1] The sense of the historic importance of the Beatles' American tours did not subside over time; if anything the hyperbole has increased steadily since that "fateful moment in the history of music," when the "most influential cultural force of the decade" arrived in the United States and "revolutionized the world of music and added a new word to the dictionary: Beatlemania."[2] The American Public Broadcasting System carried out an opinion poll in 2009 that voted the Beatles "The world's most influential artists."

No one has worked harder than EMI and Apple Corps to keep Beatlemania alive and profitable, and they have continually reissued the Beatles' records. The *Anthology* project of 1995 brought renewed interest in the band with a record set, book, and eventually multiple DVDs. A compilation of hits, *The Beatles 1*, in 2000, started off the new century with a shot of nostalgia that found its way into countless CD collections and MP3 play lists. Each re-released, repackaged set of recordings came with the fanfare a band of the Beatles' stature deserved. The Cirque du Soleil *Love* production of 2006 hailed the Beatles as "the most beloved rock band of all time" and promised a new way to experience the music. When Harmonix Music Systems introduced a computer game based on the Beatles in 2009, they proclaimed "The world's leading music game meets the greatest band in history." *The Beatles: Rock Band* was the latest version of their *Guitar Hero / Rock Band* computer game series, in which the player uses scaled-down plastic guitars to accompany the tracks played by the computer. *The Beatles: Rock Band* gave the player a chance not just to participate in the production of the music but also to experience the excitement of the Cavern Club or the Shea Stadium concert in the virtual world created by the game.

Such is the band's prestige and the evocative power of their music that it validates whatever format is used to bring it to the public. The introduction of compact discs received a lift with the release of the Beatles' catalog in digital form in 1987. When iTunes announced that the music from the "band that changed everything" was now available to download in MP3 format in 2010, the world press reported it as a major event in popular culture. Fans new and old rushed to download over 2 million songs in the first seven days, and the British Phonographic Industry proclaimed that the event "marks the coming of age of digital music."[3]

The music has endured as it has been passed down from generation to generation. Russell Richey heard his first Beatle songs when he was four, all the way back in 1970. He got the albums from an uncle and listened to them as a teenager. During his college years in the 1980s, he picked out the melodies of his favorite Beatles songs on an acoustic guitar. In the 1990s he purchased their catalog on compact discs and shared them with his family. Now, in the twenty-first century, his children are downloading the same songs to their telephones and MP3 players.[4] But each group of fans did more than just pass on the Beatles music; some of their memories and experiences were also gifted to the young. The Beatlemania generation "got married, had children and passed on the values that they learned from the Beatles and their music."[5]

Every new generation of Beatles fans receives the music within the context of the band's place in the 1960s and its role in creating the particular experience of that well-remembered decade. One of the engineers who developed *The Beatles: Rock Band* pointed out that the places, people, and spirit of the times had to be incorporated into the game because someone who just listened to the music on record was "missing out on the experience of what the Beatles meant to me." But not everyone welcomed this invocation of sixties nostalgia. The critic Lester Bangs, never a big fan of the band, commented, "I'd like to know what I have missed by not missing the Beatles," and argued that while they might have stood for the era, they have "totally exhumed it from fact" and now float aimlessly above it. Beatlemania is there to be dusted off at appropriate intervals, he said, "depending on the needs of Capitol's ledgers and our own inability to cope with the present."[6]

THE GREAT JOURNEY

Very few Beatles fans would agree with Lester Bangs' acidic assessment of the timelessness of the Beatles' music. So many have called it the soundtrack of their lives that it hardly seems worth mentioning anymore. Beatlemania made the band and their music the focus of the faithful who bought their records, went to their concerts, and hummed the tunes. For many the Beatles were at the center of their consciousness. Debbie Leavitt was far from being the only fan who admitted, "They were my whole life."[7]

Looking back at the 1960s, the fans interpreted the Beatles' narrative as part of a much larger social movement. The transition from boy band in collarless jackets to bearded hippies of the counterculture gave the Beatles a prominent position in leading the changes in music, fashions, and attitude. They were no longer just entertainers but an abstraction that represented some key elements of the sixties, especially its dreams and desires. As Paul McCartney said, the Beatles *became* the sixties to many people: "We were the symbol for everything that was happening—free love, free sex, free thinking."[8]

Critics, broadcasters, and major publications like *Time* and *Look* reinforced the widespread notion that the Beatles' music both reflected and encapsulated its time. They saw Lennon and McCartney as the F. Scott Fitzgerald or the Toulouse Lautrec of the sixties, their art capturing the essence of the decade, not just with songs that welcomed hedonism but also with more complex representations of the emptiness of the affluent society. *Sgt Pepper's Lonely Hearts Club Band* was so joyously received in the summer of 1967 because it expressed many feelings that were in the air, as Paul McCartney recognized: "It was more that it was an album for our generation. It was an album that marked the times and summed up the times."[9] Whether by fortuitous timing or brilliant planning, the Beatles' records resonated with their youthful audience as no music had ever done before, keeping pace with their emotional growth: "I was 14 in 1964, just right to GET IT; at 17 Sgt. Pepper called, She's Leaving Home, made my pilgrimage to London to meet the Fab Four." The Beatles were the pied pipers of the sixties youth movement: "They changed music, culture, the arts, society . . . and had fun leading us on a great journey."[10]

When the great journey began, the four musicians were content to listen to Brian Epstein and keep quiet on controversial issues, but their hair length and unusual dress inevitably made them representative of the younger generation. Michael Frontani has chronicled their development from lovable mop tops with a carefully controlled image to spokesmen for the youth movement. Their fame and a compliant mass media gave them the power to articulate an issue and bring it to the attention of the world. Beatlemania conferred on them unprecedented power and influence and determined that they would become the lightning rod for many controversial issues of the decade. George Harrison made the perceptive comment that everywhere the Beatles went on their American tours, "there seemed to be something going on."[11] They arrived at a time of tumultuous change. The status quo of the 1950s was being challenged by youth, African Americans, migrant workers, gays and lesbians, environmentalists, feminists, and anyone else who could make up a poster and organize a demonstration. The Cuban Missile Crisis, the standoff in Berlin, and the escalating Vietnam War were fresh in the minds of Americans when the Beatles arrived. Subsequent tours landed the band in the middle of the civil rights movement, the rise of student radicalism, and the emergence of a counterculture. Many red-button issues—drugs, racial and gender equality, foreign policy, and the role of Christianity in a materialistic world—had little to do with the Fab Four, but again, they were in the right place at the right time to address them.

Many of the important changes in politics and popular culture were driven by the growing power of youth, and it was among the young that dissent was the strongest. A country that had always claimed to have done the right thing for its children now found them in rebellion. Parents had brought up their kids under the guidance of American pediatrician Dr Benjamin Spock, who advised indulging the young and treating them as individuals. But when their children demanded the right to be heard, the parents wondered if the permissiveness advocated by Spock might have gone too far. The younger generation had strong opinions about their rights as individuals and their status in society. Imagining yourself in the middle of a golden age, and feeling that you had a part to play in this period of profound change, was one of the special emotions of the 1960s. To those who lived well in the 1960s, these were *the* times, an important turning point in history illuminated by events like Beatlemania. It did not matter if the times were not as historically significant and as happy as popularly remembered, or if it was possible that any one band's music could sum up the emotions of an entire decade—it was enough that this special generation thought it so. As the Beatles spoke to each member of those massive, screaming audiences, each of their fans felt their individual experiences were part of the great journey.

The establishment of a youth culture marked a bitterly contested debate over loud music, hair length, outlandish dress, illegal drug use, and sexual activity. The generation gap had never been wider, with accepted standards of behavior undergoing assault and sometimes transformation. Larry Kane remembered, "There was no question—immense cultural changes were underway" in the United States.[12] The Beatles were far from being radicals despite their insistence on racial equality in their audience and their opposition to war in general and the Vietnam War in particular. Many young radicals criticized their song "Revolution" as a cop out (how can the problems of society be solved by four multimillionaires telling us that things are going to be all right?), and many on the far left were disappointed that they did not use their media presence to urge for some real changes. There was little doubt who the film director Jean-Luc Godard was talking about when he said that some famous people had the means to do something positive with their media power but would not.

Yet in the conservative United States of the early sixties, growing your hair and singing about drugs (however cleverly encoded) was enough to make musicians the voice of the counterculture. It is hard to believe in retrospect, but long hair really was an explosive issue in many households, and it was associated with all sorts of subversive behavior. Some right-wing extremists like Rev. David Noebel saw Beatlemania as part of a cunning communist plot to undermine American youth: "The Beatles' ability to make teenagers weep and wail, become uncontrollable and unruly . . . is laboratory tested and approved." The conversion to Beatlemania often came at a significant moment in the politics of the American family, when the first shoots of independence blossomed into a new identity. Ken King wrote about the conflict with his mother when he became a Beatles fan in a seminar paper entitled "She Said the Beatles Killed Her Baby." To emulate the Beatles in dress, speech, or worldview was to place yourself at odds with the adult world. The mop-top hairstyle became a badge of nonconformity as well as one of the most profitable lines of Beatlemania merchandising.[13]

Rock'n'roll was moving out of the realm of hedonism and into a wider political arena when the Beatles started thinking of themselves as serious artists. Pop music now represented the ideals and aspirations of youth as well as their mindless entertainment. In the last half of the sixties, popular music from folk to hard rock became an organizing principle for youth culture. While the Beatles were singing about Lucy in the sky with diamonds, Buffalo Springfield was telling its listeners, "Stop, children, what's that sound, everybody look what's goin' down" ("For What it's Worth," 1967). The counterculture had a special place for music in its world plan, seeing it as a signifier of the power of youth to alter the society around them. Beatles fans felt that popular music could be a tangible and positive force for change, and the experiences of the civil rights movement showed how important a song like "We Will Overcome" could be in motivating people. The Beatles' status as the band that took rock'n'roll to new heights gave their opinions a great deal of weight with the younger generation. Ngaere Baxter argued that the "music [was] so good it validated their words," and Geoffrey O'Brien articulated a feeling that was widespread in the 1960s when he said, "Many came to feel that the Beatles enjoyed some kind of privileged wisdom."[14]

The fans saw the band as peers who thought like them and could be depended on to give it to them straight. They were drawn to what they perceived as the Beatles' irreverence and defiant honesty. They admired their freedom and open mindedness, which exemplified "a refreshing distrust of authority, disdain for conventions and impatience with hypocrisy."[15] The Beatles' awareness of belonging to a cohesive group with its own interests, often at odds with the establishment, made them more outspoken over time, especially after Brian Epstein was gone. Their opinions on drug use, racial discrimination, and the futility of war were broadcast globally along with the mundane details of their millionaire existence. They were now speaking out on issues not normally commented on by pop stars, and their music was being appropriated by various elements of the counterculture. Student leaders were talking about the fusion of heads, hearts, and hands as American youth grasped that its time had come, and popular music played an important part in this realization.[16]

From the mid-sixties on the Beatles' music was used not only to illuminate but also to legitimize the ideas of the counterculture. Yet there were several contradictory forces at work: the enormous wealth of the musicians, their blatant materialism, and their servitude to one of the largest capitalist organizations in the world would seem to undermine their credibility as spokesmen for radicalized youth. As usual they managed to be all things to all people. Their position as the best known entertainers of the decade gave their pronouncements more weight, but it also made them a magnet for many diverse viewpoints and causes. As Dominic Sandbrook puts it, "They had become, like cartoon characters, an elemental silhouette in which all desires and fantasies could be lived and gratified."[17]

"THE BAND THAT CHANGED EVERYTHING"

The younger generation felt that the 1960s were exceptional times and that it took exceptional lives to represent them. The fans saw the Beatles as exceptional: "They were fresh, they were new, there was just something really special and magical."[18] This sums up the appeal of the sixties: fresh, new, special, and magical are the same words baby boomers use today to remember those times. The Beatles managed to personify some of the core values of the baby boom generation: hope, change, and the opportunity for personal fulfillment. Their own rags-to-riches story had a central role in the sixties culture of optimism—anything was possible, and great things were expected of the future.

George said that the Beatles always knew that "something good was going to happen."[19] And it did. They also made millions of young people think that good things were also going to happen to them. The sixties fed off the enthusiasm for science and technology to look optimistically into the future, and the Beatles' recording career showed that embracing the new could be rewarding. There were certainly great expectations in the air when they arrived at Kennedy International Airport back in 1964. The transformative power of Beatlemania focused the expectations of a generation onto a single moment of personal enlightenment. One of them said that after watching the Beatles perform on *The Ed Sullivan Show*, "Something very important began for us that night."[20] Beatlemania provided many of the shared formative experiences that brought the sixties youth culture together.

Although the rise of the Beatles demonstrated the power and the reach of the Empires of Sound, the band articulated the rhetoric of sixties rebellion in their struggle to achieve more creative freedom from the big record company that owned their contracts. The emergence of the Mersey Sound and the British Invasion provided no examples of independent companies talking on the majors and winning in the David versus Goliath struggle that brought us rock'n'roll. Yet to many of their followers, the Beatles expressed a freedom that was a central pillar of sixties culture—not just the freedom to make their music as they wished, but the freedom that enormous wealth provides to rise above the everyday. Brian Epstein caught the essence of this when he wrote about the "big let out," the means to escape the "humdrum, dreary, god-forsaken suburbia."[21] And just like their music had risen above the banality of popular entertainment and the rigid rules of their record company, the Beatles rose above their status as pop stars.

The sixties were fundamentally the sixties because of the demographic spread of affluence. And who better to represent this than four (allegedly) working-class youths who were rich beyond anyone's wildest dreams? The pounds and dollars they earned were so many that the wealth itself played an important part in generating the superlatives that fueled Beatlemania. Even the amount of tax they paid was so large it was difficult to comprehend. There were people who cared nothing for the music and knew very little about the four musicians except the amount of money they made. The Beatles and their management stood as the outstanding example of making it in the sixties. Peter Brown tells a wonderful story about the early days of Beatlemania. When somebody walked into a pub and plonked down a large denomination bill on the counter, people would say, "Who do you think you are? Brian Epstein?"[22] The Beatles were a symbol of sixties affluence—people who could buy whatever they wanted and choose from a longer list of luxuries than ever before, from inter-room stereo systems to Greek islands. As leaders of a new, trendy materialism, they provided proof of all that was possible in the sixties.

The baby boom generation that the Beatles went on to represent, possessed great economic power, which they turned into cultural and political influence. Young people were beginning to realize that not only could they consume, they could also influence popular culture. Their choices and purchases had made the Beatles the most successful pop group ever. The frenzy of Beatlemania was an expression of the power of the teenage consumer who bought the records and magazines and then acted up in public. It was the theater of materialism in an affluent society.

The sixties were memorable for a lot of people because they were times of hope: of financial improvement, of finding love or excitement, and of changing the world and somehow making it a better place. The Beatles expressed it in their songs and lived it in their well-reported and highly regarded lives. Kathleen Stewart was part of that "generation of young Americans who simply yearned to improve upon the status quo and change the world for the better." No one inspired the baby boomers more than the Beatles. "They gave us hope," Richard Manly remembered, and it was a hope that went beyond everyday materialism. Martin Goldsmith recognized a unity of purpose with their fans: "They supplied hope and wonder and an unquenchable optimism to an age that, at best, believed deeply in the perfectibility of humankind."[23] Changing the world crops up regularly in the fans' personal interpretations of Beatlemania: "They changed EVERYTHING, artistically, socially, politically, musically, spiritually," "the Beatles changed the entire complexion of the world with their music," and "from the beginning they shaped the world with their music."[24]

It would be naive to think that any pop group could be responsible for so many significant changes. In fact, outside popular music, they influenced little of the world around them. John Lennon concluded later that the only thing the Beatles changed was fashion. The status quo remained the same, and all that optimism and hope came to nothing: "The dream is over. It's just the same only I'm thirty and a lot of people have got long hair, that's all."[25] Lennon's cynical pronouncements did little to dampen the fans' ardor, and those memories of change remain as golden and as evocative as ever. That so many people should consider the Beatles a force of social change is testament to their ability to represent change rather than initiate it. The Beatles' achievement was to embody the hopes and aspirations of a decade to their fans without doing much more than just being the Beatles.

Different audiences found different meanings in the music of the Beatles, and in some places the band did play a significant role in fostering change. In the American South they were often considered a force of communism and atheism. Yet in the home of communism and atheism, they came across as dangerous opponents rather than fellow travelers. Beatlemania in Soviet Russia had a profound effect on the liberalization of the country and accelerating the end of communism. In a documentary entitled *How the Beatles Rocked the Kremlin,* Leslie Woodhead analyzed the impact of Western pop music on Russian youth and concluded that the Beatles made the first gaping hole in the Iron Curtain, turning the young away from the communist system toward the freedom of the West. As bourgeois, decadent music, the Beatles' records were banned in Russia, which created a black market of pirated tapes—again, the tape recorder was vital in the diffusion of their music. Russian youth were so desperate to hear these recordings that they employed every means possible to duplicate them, including recording on x-ray plates. Like everybody else who heard them, Russian listeners found the Beatles new, fresh, and exciting, and the emotions stirred by the music were the joy of freedom and hope for a better future. Again, the Beatles employed perfect timing, because when their music was first heard in the Soviet Union, the patriotic afterglow of World War II had dissipated, and in its place was economic stagnation and disillusionment. Many fans across the globe described the coming of Beatlemania as a burst of color in a gray existence, and this was especially true in Russia. Woodhead argued that Beatlemania was a religion in Russia that changed the lives of millions of young people as it slowly and subtly undermined communism by attacking its soft, cultural underbelly.[26]

THE BEST OF THE BRITISH

The Beatles were perceived differently on each side of the Atlantic, standing for different things and representing different worldviews. In the United Kingdom they were considered especially English, and their achievements promoted a patriotic pride that had been aroused by the declining status of the country as a world power. To their countrymen the Beatles were a surrogate for the UK's "special relationship" with the United States, and conquering America in their effortless, lovable manner had symbolic meaning far beyond the chart successes and sold-out shows. The English interpreted this as the regaining of the imagined cultural leadership they had lost early in the century. Beatlemania had resonance for a people coming to terms with the straitened circumstances of the British Empire. What started out as mild mayhem from English fans gained meaning as it spread to the United States as part of the British Invasion of American popular culture. The terminology used ("invasion" and "conquering") has military overtones that imply a struggle for hegemony. The transatlantic entertainment industry of the nineteenth century had placed the cultural leadership in Europe, while Americans tended to lead in finance and technology. Then the development of ragtime and jazz as dance music in the early twentieth century changed the dynamics of transatlantic diffusion, wresting away cultural hegemony from Europe. The big band jazz and epic films of the 1930s and 1940s were all American. Before the Beatles, popular culture only went one way—from the United States to Europe.

What made rock'n'roll an anathema to the British establishment was not only its delinquent behavior, but also that it was another installment of inferior and debased popular culture that might be all right for dimwitted Americans but was beneath the nation that had produced Shakespeare and Milton. The British might have lost everything in the war, but they still held on to their innate feelings of cultural superiority, despite their children becoming more Americanized every day. The Beatles' success in reversing this trend, and the spectacular way it was achieved, is one of the reasons they were so popular in their home country. The empire was no more, and the roles of British ships, banks, and insurance companies in the transatlantic trade were diminished, but as the tabloids pointed out, those four Liverpool lads proved there was still some life left in the British lion.

The British press, never reluctant to wave the Union Jack and inflate national pride, started to write about the new vitality of British pop music in 1963, when papers like the *New Musical Express* noted that the English charts were no longer totally dominated by American records. The absence of American records at the top of the charts in 1963 (only Elvis managed a Number 1 that year) led papers like the *Melody Maker* to forecast that "the Beatles could take it to the Americans," a dream that had been swirling around the British entertainment industry for years. The triumphant American tour created a new Beatles story for the English press: the successful invasion that established British cultural dominance overseas.[27]

It wasn't just the Mersey Sound that had conquered America; British films, plays, and literature had also met with an enthusiastic reception in the United States. Broadway was crammed with English productions, such as *My Fair Lady*; Hollywood was looking for more irresistible British actors like Richard Burton or Peter O'Toole; and even films that dealt with the distinctly unglamorous life in the industrial North of England found an appreciative audience, both in the theaters and with the Academy of Motion Picture Arts. Americans were becoming fascinated with the British.

Walter Shenson, an American film producer based in London, obtained financial backing from (American) Columbia studios in 1958 to make a low-budget comedy called *The Mouse That Roared*, which he hoped to market in the United States. It was constructed around typically British understated humor and stereotyped characters, several of whom were played by Peter Sellers, and its plot played on the European sense of inferiority toward the United States. Its popularity with Americans showed that they could take the joke and indicated how much Anglophilia had developed there. After Columbia's unexpected success with *The Mouse That Roared*, other American film companies moved in. United Artists started operations in England with *Dr. No* in 1962, the debut of James Bond on film and a major success on both sides of the Atlantic. Just like *The Mouse That Roared*, there was a fair amount of pride and patriotism in the film, which depicted an Englishman saving the world and relegating his American colleagues as helpers rather than the dominant partners in the Cold War. Agent 007 was British to the core, and he captivated American audiences, including their president.

The unfavorable trade balance after World War II had brought the American and British film industries closer together because it created desirable financial incentives for American producers to invest in British films or set up operations in the United Kingdom. A major part of the war effort was paid for in borrowed dollars, and as European nations struggled to recover after the conflict, they found it difficult to raise the dollars needed to buy essential goods from the country that had been the arsenal of democracy. The Marshall Plan and other American initiatives to rebuild Europe recognized the problem of these large trade and currency imbalances and tried to stem the flow of currencies out of Europe. Box office receipts for American films were part of this outward flow, and governments took steps to keep these funds in Europe by restricting American studios from drawing out foreign currencies and by providing financial incentives to keep the money in Europe. An American studio could use its box office revenue in England to set up production companies in the kingdom. If it used the required number of English technicians and crew, the production would be classed as British and then benefit from incentives like the Eady Levy, which diverted tax revenue into British film companies to build up the home industry (so that it was stronger to resist American imports). Making films in England took advantage of favorable terms for American filmmakers, who could use cheap English pounds rather than expensive dollars to finance their films and then export the product to their home market. This is why *A Hard Day's Night* was made for an American company (United Artists) by an American film producer (Walter Shenson) and directed by an American (Richard Lester).

These financial arrangements benefited individuals as well as corporations. Mindful of the importance of keeping American aid workers in Europe, the IRS gave them generous tax breaks on their overseas earnings, which allowed them to keep most of this income untaxed. This not only kept aid workers in Europe but also attracted many other types of workers, such as film and television producers, photographers, designers, and journalists. The flood of American talent into the British film and television industries in the 1950s was more about generous financial incentives than the food and weather. Important film producers and some famous directors moved to England during these years, including Stanley Donen, Sidney Lumet, and Stanley Kubrick. By the time the Beatles moved to London in 1963, numerous creative Americans were already working there: film producers, directors, writers, journalists, broadcasters, and record company A&R men. All the major record companies had people in London, and several independents had representation there too. American music publishers maintained offices around Denmark Street, and American entertainment bookers kept representatives in London. U.S. newspapers and radio networks had been placing their journalists in England since the dark days of the 1940s, and the large newsgathering organizations and television networks set up London offices in the 1950s.

These influential people were the foundation for the British Invasion. Those first press and television reports about Beatlemania were sent to the United States by American journalists already working in England. The story that ran on *CBS Evening News* in December 1963 originated with one of CBS's correspondents based in London. Alexander Kendrick had read about the Beatles in the newspapers, and while covering another story, he decided to attend a performance. Impressed by the visual appeal of the fans screaming, he brought in a CBS film crew and had Josh Darsa interview the band after a concert in Bournemouth. The film was sent to CBS in New York in November but not broadcast until the national depression about President Kennedy's assassination had diminished. According to Walter Cronkite, this story aroused the interest of Ed Sullivan, who contacted him after the news and asked him for the name of the band—"those bugs or whatever they call themselves"—that he had noticed on a trip to England in search of talent. Sullivan was not doing anything differently from the impresarios who had traveled the Atlantic by sea a hundred years before, but efficient jet transportation and international phone service made the work quicker and easier.[28]

Beatlemania strengthened these ties. The Beatles' triumphant American tour increased the tempo of signing up English beat bands, and American record companies sent their scouts to Liverpool, Manchester, and London to find the next Beatles. The British Invasion had begun with the export of records but quickly moved into television programs and film created in Great Britain and distributed in the United States. American film producers and investors created a wave of English films designed for the American audience, and they were the dominant force in British film production in the late 1960s. Brian Epstein turned the Mersey Sound into a global brand. Gerry Marsden and the Pacemakers toured the United States and starred in *Ferry Cross the Mersey*, which was filmed in their hometown. Other northern acts, like Herman's Hermits, starred in several movies released in the United States by MGM. For many British Invasion bands, the American market was the reason for their existence. The British Invasion is a case study of the transatlantic reach of the Empires of Sound and of the critical importance of their skilled personnel in this exchange.

SWINGING LONDON

It should come as no surprise that the Empires of Sound were busy looking for another transatlantic marketing event as Beatlemania wound down after the band retired from touring. The same media outlets that had created Beatlemania concocted the idea of Swinging London. British newspapers and television programs took up the story, and films like *Blow Up, A Hard Day's Night,* and *Alfie* carried a vision of the excitement and diversity of the capital city to all corners of the United States. They helped form the image of Swinging London and propelled the idea that a great cultural renaissance was in full swing, encompassing music, image, style, and fashion. Influential weeklies like *Time* and *Newsweek* promoted the story in the United States and then carried it over into television and films. The American press praised the artistic achievements of Britons, such as the new National Theatre, while publishing pictures of "dolly girls" in miniskirts. Record companies quickly cashed in by promoting groups that had some slim connection to the swinging city.

Thanks to exported British theater and films, it had been cool to be English in America slightly before the Beatles arrived, but after their triumph it was even cooler. The marketing concept of Swinging London expanded the musical invasion into fashion, design, and art. The swinging sixties were coalescing into a look that originated in the fashionable shopping streets of London, where clothing designers like Mary Quant, hairdressers like Vidal Sassoon, and interior designers like Terence Conran had established themselves and were now following the Beatles into the massive American market. "I've conquered London, now I'm conquering America, next it's the world." These are the words of John Stephen, a successful clothing retailer and leading light in the Carnaby Street revival of British fashion. After establishing numerous successful boutiques in London, Stephen invaded the United States. Mary Quant and Vidal Sassoon also set up American outlets. To entrepreneurs like these, the American market was not all about money; it was also about validation.[29]

Carnaby Street, with its fashionable stores and shoppers, acted as an appropriate center for Swinging London. Easily found and photographed, it was a major tourist destination that generated millions of pounds every year. But it was not the center of Swinging London as the guidebooks maintained because it was studiously avoided by the beautiful people, a cultural elite of about two thousand very affluent young adults (in the estimation of David Bailey, the photographer who probably did the most to promote the idea). To those who knew and worked with the beautiful people, as Peter Brown did, Carnaby Street and Swinging London were no more than con games, illusions created by the tourist and fashion industries. But both these businesses were booming in 1960s England. The introduction of inexpensive jet travel had made London a major destination, and so important were the tourist dollars that the London Tourist Board was established in 1963 to encourage their flow into England. The images of Swinging London and Carnaby Street were created as part of this strategy and planted in the American media, where they flourished.

The British fashion industry was a major benefactor of Beatlemania and the British Invasion. The clothes developed by designers like Mary Quant were more than fashionable; they were designed for the times, and somehow they managed to reflect the times. Quant talked about making clothes for young people that fit into the lifestyle of "pop records and espresso bars and jazz clubs."[30] The look was all about modernity. The neat, clean, simple line of her dresses matched the short bobbed hairstyles created by Vidal Sassoon—a cut that was reproduced with the ubiquity of the Beatles' mop tops. Swinging London brought a package of British products to the United States, from special fragrances to mailorder Beatle boots. Loud pop music was essential to recreate Swinging London, and thus American musicians provided the backdrop for department stores to sell English fashions and for bars to run miniskirt contests. American businesses as diverse as barber shops and recording studios strove to copy the latest fad from London.

As leaders of fashion and the center of the highly visible but exclusive London club scene, the Beatles acted as ambassadors for Swinging London. Their favorite hangouts, like the Scotch of St James; their elaborate parties; and their happenings, like the *All You Need is Love* telecast, brought all the beautiful people together. They were part of a new British aristocracy of the 1960s. Now that they and their peers had the money to enter the English upper classes, they decided to form a new one of creativity and accomplishment. Their meritocracy recognized money, style, and achievement instead of social class. The new celebrities first came from popular music and film, but soon the new aristocracy encompassed fashion photographers, interior decorators, television producers, hair stylists, record producers, photogenic criminals, and sports stars. Celebrity was constructed in the shadow of Beatlemania, and the newly minted stars, brought to us like all the others by television, were often given names that fixed them against the example of the Beatles; there were Beatle-like comedians, hoteliers, clothes designers, footballers, and even a bull fighter who bathed in the Beatles' reflective glory.

The sixties marked the flowering of a youth culture that made it fashionable and advantageous to be young. The new fashions proclaimed, "I'm young, and I'm special," and to be young was in; to be old, out. Youth was a cult in the sixties, and its look and slang became part of the vernacular. Young England was depicted as a classless society bursting with new ideas, and the Beatles and their peers reveled in the status given to the new urban elite. Unlike the old aristocracy, they sensed changing fashions and moved with them while retaining the unswerving loyalty of their subjects. The new aristocracy represented an emerging class of artists and entrepreneurs who had demonstrated the ambition, progressiveness, and egalitarianism that everyone admired in the Americans. And they did it with style.

This freedom to rise above your social station was part of a major revision of the class system in England—a change that made an indelible impression on the 1960s. The Beatles were leading a new social elite that was admired publicly for its creativity and privately for its wealth. This unique sixties' desire to become more than you were, to rise above the ordinary, was epitomized by the transformation of four working-class lives. The baby boom generation was focused on living better than their parents and breaking away from the restrictions on personal ambition and expression that had held the previous generation in the grip of a conforming society. Although the band was singing that "nothing is real" in the final years of their career, their financial security and freedom of expression were concrete. What the Beatles revealed to their fans was a world of possibilities undreamed of before. When President Barack Obama presented Paul McCartney with the Gershwin Prize in June 2010 for services to popular music, he told him, "a grateful nation thanks you for sharing your dreams with us."[31]

The Beatles were not content to remain as pop stars, and they made it acceptable, if not desirable, to have active interests in film, literature, and art, which they hoped would overshadow their roles as entertainers. The singer Marianne Faithful, one of the beautiful people, noted that "all these different métiers—film, painting, sculpture, rock'n'roll, classical music, everything, photography—were opening up and you could switch, you could go from one to another."[32] In the boundless optimism of the sixties, the Beatles rewrote the rules for rock'n'roll stardom as they moved away from the "boys" cultivated by Brian Epstein to self-conscious artists who flirted with the avant-garde, and then to businessmen who planned to take advantage of all the technological advances in entertainment. When asked about the Beatles' future, John and Paul described it in terms of films and music. John explained that their new business venture of Apple Corps in 1968 was "going to be records, films, and electronics . . . which all tie up."[33] The Beatles were important in the construction of the 1960s because they successfully married teen music with bohemian culture, merging artistic ambition with pop sensibility, and in doing so, they carried forward the torch of the swinging sixties.

THIS IS HISTORY

Beatlemania has assumed heroic significance in the memories of the hundreds of thousands of Americans who experienced it. As one of them said, "Other than the birth of my children, it was the single most important highlight of my life." Fifty year later the fans have not forgotten the exhilaration of Beatlemania: "It's like it happened yesterday," and "I'll never forget that as long as I live." As with the other monumental events of the 1960s, a person who lived through Beatlemania can say, "[I remember] exactly where I was the first time I heard 'I Want to Hold Your Hand.' "[34] The scrapbooks and mementos have all been saved; the outfits worn during those historic concerts have been preserved in plastic; and most of the treasured records have survived, including those too worn out to be played. Nobody seems to have thrown away the ticket stub from a Beatles concert, and some stubs have ended up in bank safe deposits. Beatlemania has been recorded on tapes, photographs, movie film, clothes, and in the fabric of many other artifacts. For the people who experienced Beatlemania, it remains a treasured memory. As a young lady said of a Beatles concert in San Francisco, the experience of "screaming like a banshee" fresh in her mind, "You knew you were in the presence of greatness . . . This is gonna go down in history."[35]

That these historic events should be played out in the United States surprised no one, which is why Beatlemania has its unique power; it would not have impressed that many people if it had been restricted to the United Kingdom or Australia. The rise of the Beatles represented the American dream writ large. As Brian Epstein used to say, "What could be more important than the United States?" And of course there was nothing. The culture of superlatives that filled the narrative of Beatlemania was only possible in the country that already had it all and appeared larger than life to the rest of the world. If you are going to be the best and the biggest, there was only one place in the sixties where you could do this—this was the view of Brian Epstein and Paul McCartney, who said that the acclaim of America was the "definitive big flash, the clincher."[36] In many ways the excesses of Beatlemania were a compliment to their American fans, a signifier of their attainment of sixties cool and their crowning achievement of participating in it.

Attending a concert was the ultimate act of fandom that placed you in the epicenter of an overwhelming decade: "I was so caught up in this moment, the reality was just being there was the thrill."[37] It was the feeling that this was an important moment, a critical juncture in both your own life and the sixties as a whole, which made documenting it so important. The cameras and tape recorders did more than preserve the moment; they also justified its importance. One person who attended the Shea Stadium concert concluded that this was "more than rock'n'roll. This is history."[38]

Fans did not just passively consume Beatlemania as part of a transaction with the capitalist entertainment industry; they integrated it into their life to help define who they were and what they hoped to achieve. Whether it was Charles Manson or the person sitting next to you at the concert, every fan read something personal and significant into the Beatles' music, for better or worse.[39] The music affirmed many individual quests for self-fulfillment and provided validation for social movements both real and imagined. Beatlemania made show business history on one level, but it endures in our collective memories because it gave teenagers the chance to be a part of what they considered a significant period in history. Geri Montefusco was at the concert the Beatles gave at Shea Stadium, in New York in 1965: "Yes, the Beatles changed the World, and continued to change it until they broke up!"[40]

Beatlemania announced the coming of age of the baby boom generation at the peak of its purchasing power and optimism. The Beatles underlined the moment of significant social change to millions of young people, as a fan pointed out: "Their appearance gave us our first sense of youth as power." But no matter how much the world was changing, the most important changes were those the Beatles brought to individual lives. Their example was so powerful on the young:

"The Beatles were living large when I was small, and it struck me that I might do better by going out into the world, which it seemed to me they were doing in a very big way." Anthony DeCurtis sums up this feeling when he argues that some bands change your life, but more important are the bands that shape your life and make you the person you are.[41]

Beatlemania provided a teenage vision of the joy of life to millions, illuminating humdrum lives with the prospect of happiness and personal fulfillment. Perhaps it was the conductor and composer Leopold Stokowski who put his finger on the root cause of Beatlemania when he said that young people were "looking for something in life that can't always be found . . . we are all looking for the vision of ecstasy of life. I am too." At its core Beatlemania was an expression of joy. In a *New York Times* magazine article, Frederick Lewis concluded, "To see a Beatle is joy, to touch one paradise on earth." Millions of the fans would agree with this: "They gave me such a feeling of happiness . . . they put a spell on me that has never been broken."[42]

Notes

1. Tony Bramwell claims that the two knew of one another well before the meeting at the church fete; *Magical Mystery Tours: My Life with the Beatles* (London: Portico, 2005), 15. Paul's house in Allerton was only a short distance from John's, and St Peters church was situated between the two; Ron Jones, *The Beatles' Liverpool* (Liverpool: Ron Jones, 2009), 50.

2. Bramwell, *Magical Mystery Tours*, 2.

3. Jann Wenner, ed., *Lennon Remembers* (New York: Popular Library, 1971), 14. Paul McCartney proudly remembered having "a very diverse little record collection"; Bob Spitz, *The Beatles* (New York: Little Brown, 2005), 111. Deborah Geller, ed., and Anthony Wall, *The Brian Epstein Story* (London: Faber, 1999), 9.

4. Unless otherwise noted, all Beatle quotes come from the *Anthology* DVD set (Apple, 2003). Bill Harry, *Liverpool: Bigger than the Beatles* (Liverpool: Trinity Mirror NW, 2009), 54. "Heartbreak Hotel" entered the British charts in May 1956.

5. Paul McCartney said that the Beatles "didn't stop playing" *The Freewheelin' Bob Dylan* album during the three weeks they spent in Paris in January 1964; Spitz, *The Beatles*, 533. John Lennon was especially influenced by Dylan's songwriting; see Walter Everett, *The Beatles as Musicians: The Quarry Men through Rubber Soul* (New York: Oxford, 2001), 255. Bob Dylan was equally enthusiastic about the Beatles: "I knew they were pointing the direction that music had to go"; quoted in Tim Riley, *Tell Me Why: The Beatles: Album by Album, Song by Song, the Sixties and After* (New York: Da Capo, 2002), 69.

6. *The Beatles Anthology* (San Francisco: Chronicle, 2000), 116; Geller and Wall, *Epstein Story*, 67.

7. P. Willis-Pitts, *Liverpool The Fifth Beatle: An African American Odyssey* (Littleton, CO: Amozen, 2000), 37.

8. Geller and Wall, *Epstein Story*, 51.

9. Bramwell, *Magical Mystery Tours*, 32.

10. Quoted from *Melody Maker*, 31 Oct. 1964, in W. Fraser Sandercombe, *The Beatles Press Reports* (Burlington, Canada: Collector's Guide, 2007), 98.

11. Geller and Wall, *Epstein Story*, 30.

12. *Daily Mirror*, 22 Oct. 1963, 16–17; John Blaney, *Beatles for Sale* (London: Jawbone Press, 2008), 67.

13. George Harrison replied to this telegram with "Please order four new guitars"; Everett, *Beatles as Musicians*, 117. There is some debate among the surviving participants and Beatle scholars whether this was an audition or a bona fide session. See Paul McCartney interview in Mark Lewisohn, *The Beatles Recording Sessions: The Official Abbey Road Studio Session Notes* (New York: Harmony, 1988), 6; Bramwell, *Magical Mystery Tours*, 65–66.

14. The concept of a Top 10 list depends on the credibility of its data, and there are many different charts and several different ways of counting sales. There was never any one single Top 10 as implied in this book and many others. The lists were compiled from several different sources. In Great Britain the leading music papers, *Melody Maker*, *New Musical Express*, *Disc*, and *Record Mirror*, all published charts, and the Top 10 we refer to is an amalgamation of these lists tempered a little by time, hindsight, and convenience. American Decca first noticed "My Bonnie" and released it early in 1962, followed by MGM; Bruce Spizer, *The Beatles Are Coming: The Birth of Beatlemania in America* (New Orleans: 498 Productions, 2004), 91. At first the Beatles were identified on the label as "The Beat Brothers."

15. Harold Montgomery quoted in Garry Berman, *"We're Going to See the Beatles!" An Oral History of Beatlemania* (Santa Monica, CA: Santa Monica Press, 2008), 58.

16. Deborah McDermott quoted in ibid., 86–87.

17. Spizer, *Beatlemania*, 101. This was the first Beatle LP issued in the United States, by Vee Jay in January 1964.

18. Ali Catterall and Simon Wells, *Your Face Here: British Cult Movies Since the Sixties* (London: Fourth Estate, 2002), 1.

19. Lee Ballinger, *Lynyrd Skynyrd: An Oral History* (New York: Spike, 2002), 108–9.

20. Robert Freeman tells the story of his work with the Beatles in the sumptuously illustrated *The Beatles: A Private View* (New York: Big Tent, 2003). Ken Sharp, "Astrid Kirchherr," interview in Sean Egan, ed., *The Mammoth Book of the Beatles* (Philadelphia: Running Press, 2009), 428. The photographer who took the most pictures of the Beatles from 1962 to 1964 was Dezo Hoffman, *With the Beatles: The Historic Photographs of Dezo Hoffman* (London: Omnibus, 1982).

21. Dave Marsh, *The Beatles Second Album* (New York: Rodale, 2007), 85.

22. Charles Tillinghurst, *How Capitol Got the Beatles* (Denver: Outskirts, 2008), 60.

23. Michele S. from Shinagawa, Tokyo, posting in the customer comment section of the Amazon.com page devoted to *Beatles: The Capitol Albums, Vol. 1*.

24. George Martin was not impressed with their musicianship or their songs, thinking "their songwriting ability had no saleable future," Everett, *Beatles as Musicians*, 118.

25. Tillinghurst, *How Capitol Got the Beatles*, 7.

26. Geller and Wall, *Epstein Story*, 71; Tillinghurst, *How Capitol Got the Beatles*, 9.

27. Spizer, *Beatlemania*, 16.

28. Spitz, *The Beatles*, 441. George Martin said that somebody at EMI could have pulled rank on the Capitol management, but nobody did; *All You Need Is Ears* (New York: St. Martin's 1979), 160. Gary Marmorstein claims it was a New York–based Capitol executive,

Manny Kellum, who finally persuaded the Los Angeles managers to issue a Beatles record; *The Label: The Story of Columbia Records* (New York: Thunder's Mouth, 2007), 443.

29. Geoff Emerick, *Here, There and Everywhere: My Life Recording the Music of the Beatles* (New York: Gotham, 2007), 73.

30. Spizer, *Beatlemania*, 74.

31. Michael R. Frontani, *The Beatles: Image and the Media* (Jackson: University Press of Mississippi, 2007), 23.

32. Spizer, *Beatlemania*, 73.

33. Barry Miles, *The Beatles: A Diary* (London: Omnibus, 2002), 101. This was especially good news because the single had just been knocked out of the Number 1 spot in the UK charts by the Dave Clark Five.

34. Kenneth Womack, *Long and Winding Roads: The Evolving Artistry of the Beatles* (New York: Continuum, 2007), 79. See also Spizer, *Beatlemania*, 115.

35. Brian Epstein, *A Cellar Full of Noise* (New York: Pocket Books, 1967), 60.

CHAPTER 2. BEATLEMANIA

1. Peter Brown, *The Love You Make: An Insider's Story of the Beatles* (New York: McGraw-Hill, 1983), 119; *Daily Mirror* 8 Feb. 1964, 1.

2. This was the view of the *Melody Maker*, a leading English music paper, on 26 Sept. 1964, reproduced in W. Fraser Sandercombe, *The Beatles Press Reports* (Burlington, Canada: Collector's Guide, 2007), 91; Virginia Maita quoted in www.thirteen.org/beatles/the-beatles.

3. Barry Miles, *The British Invasion: The Music, The Times, The Era* (New York: Sterling, 2009), 102.

4. *Time*, 15 Nov. 1963, 64; Kevin Howlett, *The Beatles at the Beeb: The Story of Their Radio Career* (London: BBC, 1982), 33–34; Sam Leach, *The Rocking City: The Explosive Birth of the Beatles* (Merseyside: Pharaoh Press, 1994), 130.

5. Bob Spitz, *The Beatles* (New York: Little Brown, 2005), 521; *San Francisco Examiner*, 19 Aug. 1964, quoted in Barry Miles, *British Invasion*, 105.

6. Larry Kane, *Ticket to Ride* (Philadelphia: Running Press, 2003), 86.

7. Martin Goldsmith, *The Beatles Come to America* (New York: John Wiley and Sons, 2004), 133; Penny Wagner quoted in Berman, *"We're Going to See the Beatles!" An Oral History of Beatlemania* (Santa Monica, CA.: Santa Monica Press, 2008), 129.

8. *Daily Mirror*, 10 Sept. 1963, 12–13; JoAnne McCormack quoted in Berman, *See the Beatles*, 121.

9. Claire Krusch, quoted in Berman, *See the Beatles*, 124.

10. Maggie Welch quoted in Berman, *See the Beatles*, 117; George Harrison said in *The Beatles Explosion* (Legend Films, 2008), DVD, that "two tons a night" were thrown at them.

11. Ron Sweed quoted in Dave Schwensen, *The Beatles in Cleveland* (Vermilion, OH: North Shore Publishing, 2007), 48.

12. Kane, *Ticket to Ride*, 49.

13. Philip Norman, *Shout: The Beatles and Their Generation* (New York: Simon and Schuster, 1981), 240.

14. Kane, *Ticket to Ride*, 35; Spitz, *The Beatles*, 522.

15. Tony Bramwell, *Magical Mystery Tours: My Life with the Beatles* (London: Portico, 2005),, 90–91.

16. *New York Times*, 29 Aug. 1964, 6.

17. Bramwell, *Magical Mystery Tours*, 76.

18. Spizer, *The Beatles Are Coming: The Birth of Beatlemania in America* (New Orleans: 498 Productions, 2004), 70, 68, 56; *Newsweek*, 18 Nov. 1963, 104.

19. *Newsweek*, 18 Nov. 1963, 64; *Daily Mirror*, 22 Oct. 1963, 16–17.

20. Paul Johnson, "The Menace of Beatlism," *New Statesman*, 28 Feb., 1964, in June Skinner Sawyers, ed., *Read the Beatles* (New York: Penguin, 2006), 53; Theodor Adorno, *The Culture Industry* (London: Routledge, 1981), 98–101.

21. Julia Baird in *Long and Winding Road* (Koch International, 2003), DVD; Geoff Emerick, *Here, There and Everywhere: My Life Recording the Music of the Beatles* (New York: Gotham, 2007), 63.

22. Goldsmith, *Come to America*, 63.

23. *Mersey Beat*, 11 April 1963, 7. This article is part of Bob Wooler's retrospective history of the Beatles.

24. *Daily Mirror* 14 Oct. 1963, 2.

25. *Daily Mail*, 21 Oct. 1963, 1, 3.

26. Jonathan Gould, *Can't Buy Me Love: The Beatles, Britain, and America* (New York: Harmony, 2007), 164; Goldsmith, *Come to America*, 96; *Daily Mirror*, 6 Nov. 1963, 6.

27. Goldsmith, *Come to America*, 115.

28. *Time*, 15 Nov. 1963, 64; *Newsweek*, 18 Nov. 1963, 104; Spizer, *Beatlemania*, 88.

29. Barry Miles, *The Beatles: A Diary* (London: Omnibus, 2002), 106.

30. Spitz, *The Beatles*, 520; Goldsmith, *Come to America*, 154; Miles, *British Invasion*, 111.

31. Ibid., 462.

32. Michael R. Frontani, *The Beatles: Image and the Media* (Jackson: University Press of Mississippi, 2007), 31.

33. Charles Pfeiffer, Carol Cox, Betty Taucher, and Penny Wagner quoted in Berman, *See the Beatles*, 73–75; Jay Willoughby interview by Jay Dismukes, 1996, Beam Oral History Project, Birmingham, Alabama.

34. Betty Taucher, Shaun Weiss quoted in Berman, *See the Beatles*, 78.

35. Spizer, *Beatlemania*, 193.

36. Brian Epstein in the *Melody Maker*, quoted in Sandercombe, *Press Reports*, 84; Bernstein's account is in his foreword to Berman, *See the Beatles*, 8–10.

37. *Daily Mirror* 20 Jan. 1964, 7.

38. From *Record Mirror*, *Mersey Beat*, and *Disc*, quoted in Sandercombe, *Press Reports*, 44, 80, 143.

39. Jerry Bishop quoted in Schwensen, *Beatles in Cleveland*, 57.

40. Geller and Wall, *Epstein Story*, 50; Paul D. Mertz posted on www.thirteen.org/beatles/the-beatles.

41. Spitz, *The Beatles*, 397. George Melly went on to write an important book about the effects of pop music on British culture; *Revolt into Style* (London: Allen Lane, 1970).

42. Steven D. Stark, *Meet the Beatles* (New York: Harper, 2005), 15; June Harvey and Debbie Levitt quoted in Berman, *See the Beatles*, 59, 69.

43. From *Disc and Music Echo*, 11 Nov. 1967, quoted in Sandercombe, *Press Reports*, 214, 218.

44. Frontani, *Image and Media*, 38.

45. *New York Times*, 17 Feb. 1964, 1, 20.

46. Ibid., 23 Feb. 1964, 15, 69–70.

47. Lennon quote from *Long and Winding Road* (DVD).

48. Goldsmith, *Come to America*, 4, 3.

49. Gould, *Can't Buy Me Love*, 100.

50. Ibid., 217–18; Stark, *Meet the Beatles*, 32.

51. Posted on www.thirteen.org/beatles/the-beatles.

52. Lester Bangs, "The Withering Away of the Beatles," in Egan, *Mammoth Beatles*, 352.

53. Ian Inglis, "The Beatles Are Coming! Conjecture and Conviction in the Myth of Kennedy, America and the Beatles," *Popular Music and Society* 24, no. 2 (Summer 2000): 93–108.

54. Spizer, *Beatlemania*, 64.

55. Cathy McCoy-Morgan quoted in Berman, *See the Beatles*, 51; Susan Hanrahan quoted in www.thirteen.org/beatles/the-beatles.

56. June Harvey quoted in Berman, *See the Beatles*, 69.

CHAPTER 3. LIVERPOOL

1. Spitz, *The Beatles*, 458.

2. Francois Vigier, *Change and Apathy: Liverpool and Manchester during the Industrial Revolution* (Cambridge: MIT Press, 1970), 48.

3. Ron Jones, *The American Connection* (Whirral, Lancs.: R. Jones, 1986), 14.

4. Vigier, *Change and Apathy*, 184.

5. J. Ramsay Muir, *A History of Liverpool* (Liverpool: University of Liverpool Press, 1907), 305.

6. Harry, *Liverpool*, 22.

7. R. J. Broadbent, *Annals of the Liverpool Stage* (Liverpool: Edward Howell, 1908), 338.

8. Ibid., 21, 100–101, 358; Peter McCauley, *Music Hall in Merseyside* (Burton on Trent: Overseas Press, 1982), 19.

9. Broadbent, *Liverpool Stage*, 219, 100–101, 189.

10. Louis S. Warren, *Buffalo Bill's America: William Cody and the Wild West Show* (New York: Knopf, 2005), 282–87.

11. Broadbent, *Liverpool Stage*, 282.

12. Ibid., 51.

13. David Robinson, *Chaplin: His Life and Art* (New York: McGraw-Hill, 1985), 77.

14. Geoffrey O'Brien, "Seven Fat Years," in Sawyers, *Read the Beatles*, 173.

15. Kenneth Womack argues for "the powerful effects of nostalgia as the group's fundamental literary and musical métier," *Long and Winding Roads*, 54; George Melly said the Beatles "were happiest when celebrating the past," *Revolt into Style*, 115. You only have to watch the Beatles on television variety shows like *Morecambe and Wise* to appreciate the ease with which all four of them could work through a vaudeville standard like "On Moonlight Bay"; *Anthology* (DVD).

CHAPTER 4. THE PROMISED LAND

1. Peter Hennessey, *Having It So Good: Britain in the Fifties* (London: Allen Lane, 2006), 19.

2. Paul Day, *The Rickenbacker Book* (San Francisco: Miller Freeman, 1994), 29.

3. Pete Frame, *The Restless Generation* (London: Rogan House, 2007), 154.

4. Gould, *Can't Buy Me Love*, 25.

5. Martin, *All You Need Is Ears*, 158; Geller and Wall, *Epstein Story*, 66.

6. Harry, *Liverpool*, 107.

7. Bramwell, *Magical Mystery Tours*, 36.

8. Steven Watts, *The Magic Kingdom: Walt Disney and the American Way of Life* (New York: Houghton Mifflin, 1997), 314–15.

9. Advertising copy for *Don't Knock the Rock* (Columbia Pictures, 1956).

10. Hennessey, *Having It So Good*, 491. Gillian Shephard later became a cabinet minister in a Conservative government.

11. Brian Southall, *Abbey Road* (Wellingborough, UK: Patrick Stevens, 1982), 55.

12. N. E. Fulcanwright, liner notes for *Brylcreemed Boys and Beehived Girls* (Acrobat Music, 2008).

13. Cynthia Lennon, *A Twist of Lennon* (London: W.H. Allen, 1978), 15.

14. Frame, *Restless Generation*, 125.

15. Geller and Wall, *Epstein Story*, 41–42.

16. Jurgen Vollmer, *The Beatles in Hamburg* (Munich: Schirmer/Mosel, 2004), 6.

17. Ibid., 5.

18. Neil Aspinall quoted in *Anthology*, 68.

19. Kevin McManus, *Nashville of the North* (Liverpool: Institute of Popular Music, 1996), 1.

20. Norman, *Shout*, 67.

21. Harry, *Liverpool*, 44; McManus, *Nashville of the North*, 1.

22. Willis-Pitts, *The Fifth Beatle*, 2.

CHAPTER 5. SKIFFLE

1. Harry, *Liverpool*, 43–44.

2. Ibid., 157.

3. McManus, *Nashville of the North*, 7.

4. Harry, *Liverpool*, 43–44, 160

5. McManus, *Nashville of the North*, 2.

6. Tricia Jenkins, *Let's Go Dancing: Dance Band Memories of 1930s Liverpool* (Liverpool: Institute of Popular Music, 1994), 37; Harry, *Liverpool*, 130–32.

7. Harry, *Liverpool*, 45.

8. McManus, *Nashville of the North*, 7–8.

9. Harry, *Liverpool*, 47.

10. Jones, *American Connection*, 8.

11. Wenner, *Lennon Remembers*, 185.

12. Kane, *Ticket to Ride*, 258.

13. Wenner, *Lennon Remembers*, 184.

14. Willis-Pitts, *The Fifth Beatle*, 40

15. This was the judgment of BBC radio producer Peter Pilbeam; Howlett, *The Beatles at the Beeb*, 10.

16. *Mersey Beat*, 28 March 1963, 11; Willis-Pitts, *The Fifth Beatle*, 41.

17. Hank Walters quoted in McManus, *Nashville of the North*, 11; Leach, *Rocking City*, 33.

18. Frame, *Restless Generation*, 3.

19. Ibid., 20.

20. Harry, *Liverpool*, 49.

21. Andy Babiuk, *Beatles Gear* (London: Backbeat, 2002), 10.

22. Harry, *Liverpool*, 47–48.

23. McManus, *Nashville of the North*, 11.

24. Martin, *All You Need Is Ears*, 59; Frame, *Restless Generation*, 207.

25. Lennon, *Twist of Lennon*, 35.

26. Harry, *Liverpool*, 59–60; Brown, *The Love You Make*, 35. Ray O'Brian has published two volumes of *There Are Places I'll Remember* that list the venues the Beatles played.

CHAPTER 6. ROCK'N'ROLL COMES TO BRITAIN

1. Some small labels were set up by artist management to distribute recordings made by their clientele. For example, Polygon Records was established in 1949 by the father of singer Petula Clark and was quickly gobbled up in 1955 by Pye, a leading manufacturer of radios, televisions, and electronics.

2. Oriole recorded many of the beat groups who failed to make a connection with a major company in London, but it did so after the event and had no role in creating the Mersey Sound. One of the first Liverpool groups to get a record contract was Derry and the Seniors, followed by the Blue Mountain Boys, who in 1962 released "Drop Me Gently" on the Oriole label; McManus, *Nashville of the North*, 31.

3. Charlie Gillett, *The Sound of the City* (New York: Pantheon, 1983), 70.

4. Gould, *Can't Buy Me Love*, 121. George Martin places the blame squarely on the chairman, Sir Earnest Fisk; *All You Need Is Ears*, 40–41. When Columbia ended its licensing agreements with EMI in 1953, Philips was waiting in the wings. One of the largest

electrical manufacturers in Europe, Philips was attracted to Columbia's catalog because two of its biggest stars, Frank Sinatra and Doris Day, were also movie stars whose films were very popular in Europe; Marmorstein, *The Label*, 205.

5. Frame, *Restless Generation*, 317.

6. In 1974 Jim Dawson sent a questionnaire to John Lennon, which Lennon filled in and returned. At the bottom of the sheet he wrote, "I WAS Buddy Holly!" See everything2 .com/index.pl?node_id=979603; Frame, *Restless Generation*, 285.

7. Joe Meek was a highly innovative recording engineer and an astute businessman who could "smell" a hit record. His Triumph and RGM labels were among the few independents in the UK that dealt in youth music and developed new sounds. Nevertheless, he rarely strayed from the American pop paradigm and dismissed the Beatles' records as derivative when they first came out.

8. Harry, *Liverpool*, 39–40.

9. Wenner, *Lennon Remembers*, 184; Frame, *Restless Generation*, 271.

10. *Mersey Beat*, 28 March 1963, 2.

11. McManus, *Nashville of the North*, 4.

12. Frame, *Restless Generation*, 334.

13. *Mersey Beat*, 20 June 1963, 2.

14. Frame, *Restless Generation*, 437; Geller and Wall, *Epstein Story*, 38.

15. Cliff and the Shadows made movie after movie with virtually the same plot, like *The Young Ones* (1961). Twenty-five years later, the Shadows' safe, squeaky clean, bland entertainment was satirized in an English comedy show with the same name as the film.

16. Harry, *Liverpool*, 80.

17. Maureen Cleave, "Why the Beatles Create All that Frenzy," *Evening Standard*, 2 Feb. 1963, in Egan, *Mammoth Beatles*, 25.

18. Bramwell, *Magical Mystery Tours*, 3.

19. *Mersey Beat*, 20 June 1963, 8; Lennon, *Twist of Lennon*, 36.

20. *Mersey Beat*, 11 April 1963, 7.

21. Miles, *Diary*, 31.

22. Emerick, *Here, There and Everywhere*, 59; Gillett, *Sound of the City*, 263.

23. Janet Lessard, quoted in Berman, *See the Beatles*, 50.

24. The Beatles played with Bruce Channel at the Tower Ballroom, Liverpool, on 21 June 1962. Considering "Hey Baby" came out in March, John Lennon probably learned it from Channel's harmonica player, Delbert McClinton, that night; Harry, *Liverpool*, 77.

25. Carol Moore, quoted in Berman, *See the Beatles*, 39; *New York Times*, 17 Feb. 1963, 1.

26. Maureen Cleave, "All that Frenzy," in Egan, *Mammoth Beatles*, 24.

27. *Newsweek*, 18 Nov. 1963, 104.

CHAPTER 7. THE LOOK

1. Geller and Wall, *Epstein Story*, 57.

2. *Anthology*, 48.

3. Barbara Allen quoted in Berman, *See the Beatles*, 48.

4. Emerick, *Here, There and Everywhere*, 41.

5. *Anthology*, 77; Vollmer, *Beatles in Hamburg*, 14.

6. Geller and Wall, *Epstein Story*, 72.

7. Epstein changed Billy's name from Ashton because he felt it was "too posh." Alistair Taylor recalls that Epstein knew that Kramer could not sing but thought he had just the right image for a pop singer; Spitz, *The Beatles*, 368. Epstein would not let him appear on *Juke Box Jury*, because it would require a lot of talking, and he decided that Kramer's diction was poor and that he needed elocution lessons; Geller and Wall, *Epstein Story*, 105.

8. Epstein first asked Peter Kaye photographers to use a studio to shoot the band, but the results were flat and disappointing, so Bill Connell next photographed them on a dump site called The Bally. Interview with Margaret Roberts, March 2011; Gareth Pawlowski, *How They Became the Beatles* (New York: Dutton, 1989), 81.

9. Babiuk, *Beatles Gear*, 28

10. Brian Southall with Julian Lennon, *Beatles Memorabilia: The Julian Lennon Collection* (London: Goldman, 2010), 17.

11. Sam Leach remembered how bad the Beatles' amps were; *Mersey Beat*, 4 Oct. 1962, 4. Comments of Norman Smith, Abbey Road engineer, in Babiuk, *Beatles Gear*, 60. The Vox story is told in Babiuk, *Beatles Gear*, 81.

12. Frontani, *Image and Media*, 45.

13. Epstein added this significant qualifier: "I was excited about their prospects, provided some things could be changed"; Babiuk, *Beatles Gear*, 59.

14. This was the advertising copy for Hunter Davies, *The Beatles: The Authorized Biography* (New York: McGraw-Hill, 1968).

15. Interview with George Harrison, *Melody Maker*, 7 Nov. 1964, in Sandercombe, *Press Reports*, 99.

16. *Mersey Beat*, 13 Feb. 1964, in ibid., 41.

17. Schwensen, *Beatles in Cleveland*, 17; John Penuel interviewed by James K. Turnipseed, Nov. 2009.

18. Frontani, *Image and Media*, 61.

19. Both John and Paul were brought up in households where speaking "properly" was highly prized. John's de facto mother, Mimi Stanley Smith, was horrified by his "wacker" accent when she saw him on television; Philip Norman, *John Lennon: The Life* (New York: Harper Collins, 2008), 303. Letter to *Mersey Beat*, 7 Aug. 1961, in Sandercombe, *Press Reports*, 7.

20. Carol Cox quoted in Berman, *See the Beatles*, 182.

21. *New Record Mirror*, 19 Oct. 1963, in Sandercombe, *Press Reports*, 25.

22. *Mersey Beat*, 1 Nov. 1962, in ibid., 12.

23. Ibid., 12 Sept. 1963, 9.

24. Geller and Wall, *Epstein Story*, 72.

25. *Disc Weekly*, 12 Sept. 1964, in Sandercombe, *Press Reports*, 87; fans shown on *Imagine* (Warner, 1988), DVD. Apple Corps was formed by the Beatles in 1967 to handle their business affairs after the death of Brian Epstein.

26. Wenner, *Lennon Remembers*, 12; Brown, *The Love You Make*, 146–47.

27. Gould, *Can't Buy Me Love*, 183–84.

28. Derek Taylor, liner notes for *The Beatles: Live at the BBC* (EMI Records, 1994), 4.

29. Geller and Wall, *Epstein Story*, 23.

30. As historian Erik Barnouw has pointed out, television drew on the programs, business strategies, and institutions of radio. It evolved from a "radio industry born under military influence and reared by big business"; *Tube of Plenty: The Evolution of American Television* (New York: Oxford University Press, 1990), 112.

31. Leach, *Rocking City*, 138.

32. "Oh Boy!" *BFI Screenonline*, www.screenonline.org.uk/tv/id/561801/index.html; "Six-Five Special," *BFI Screenonline*, www.screenonline.org.uk/tv/id/561782/index.html; "Good, Jack," *BFI Screenonline*, www.screenonline.org.uk/people/id/574989/index.html.

33. "Oh Boy!," *Television Heaven*, Televisionheaven.co.uk/ohboy.htm.

34. Blaney, *Beatles for Sale*, 53.

35. Goldsmith, *Come to America*, 86.

36. *Melody Maker*, 25 Dec. 1965, in Sandercombe, *Press Reports*, 151.

37. Bob Neaverson, *The Beatles Movies* (London: Cassell, 1997), 27.

38. Andrew Sarris, "A Hard Day's Night," *Village Voice*, 27 Aug. 1964, in Sawyers, *Read the Beatles*, 56–58.

39. *Long and Winding Road* (DVD); Catterall and Wells, *Your Face Here*, 13.

40. Catterall and Wells, *Your Face Here*, 6.

41. Roy Armes, *A Critical History of the British Cinema* (New York: Oxford University Press, 1978), 259.

42. Catterall and Wells, *Your Face Here*, 10.

43. Frontani, *Image and Media*, 130.

CHAPTER 8. THE FANS

1. *Daily Mail*, 1 Jan. 1964, 3.

2. Ibid., 22 Feb. 1964, 1.

3. Gloria Steinem, "Beatle with a Future," *Cosmopolitan*, Dec. 1963, in Sawyers, *Read the Beatles*, 62.

4. Frontani, *Image and Media*, 93. You could buy a Beatle Krunch Ice Cream Bar and put it in a Beatle lunchbox. There were also Beatle toiletries, hair products, sneakers, board games, and record players.

5. Brian Braithwaite and Joan Barrett, *The Business of Women's Magazines* (London: Associated Business Press,1970), 30–37; David Galassie, "Gloria Stavers and 16 Magazine," www.loti.com/sixties_history/Gloria_Stavers_and_16_Magazine.htm; "65 Years of Seventeen!" *Seventeen*, www.seventeen.com/fun/articles/65th-anniversary-cover-archive?click=img_sr.

6. Norman, *John Lennon*, 323–24. On the other hand, an unfavorable story in these magazines, such as John's "bigger than Jesus" interview, reprinted in *Dateline*, could do a lot of damage to the band's image.

7. *Daily Mirror*, 14 Oct. 1963, 2.

8. *Disc and Music Echo*, 2 July 1966, in Sandercombe, *Press Reports*, 167.

9. Harold Montgomery quoted in Berman, *See the Beatles*, 205.

10. Stark, *Meet the Beatles*, 110.

11. *Mersey Beat*, 13 Dec. 1962, 3.

12. Pawlowski, *How They Became Beatles*, 64, 80.

13. Blaney, *Beatles for Sale*, 171.

14. These backstage meetings were described as "sacred events" by Kane, *Ticket to Ride*, 121.

15. Kane, *Ticket to Ride*, 86; *Disc and Music Echo*, 2 July 1966 in Sandercombe, *Press Reports*, 168.

16. Womack, *Long and Winding Roads*, 64.

17. Ibid., 68–69, 55.

18. Any citation with Christian name and first initial, such as Amy B., comes from Bill Adler, ed., *Love Letters to the Beatles* (New York: Putnam's, 1964), unpaginated.

19. Betty Taucher quoted in Berman, *See the Beatles*, 197.

20. Linda Cooper quoted in ibid., 46.

21. Linda Cooper, Maryanne Laffite, and Mary Ann Collins quoted in Berman, *See the Beatles*, 74, 138; Martin, *All You Need Is Ears*, 161.

22. Kane, *Ticket to Ride*, 120.

23. *New York Times*, 23 Feb. 1964, 15, 69–70.

24. Janet Lessard and Barbara Allen quoted in Berman, *See the Beatles*, 74, 45,127; Steinem, "Beatle with a Future," 62.

25. Elizabeth Hess, "The Women," *Village Voice*, 8 Nov. 1994, 91; Carol Moore quoted in Berman, *See the Beatles*, 145.

26. Barbara Ehrenreich, Elizabeth Hess, and Gloria Jacobs, "Girls Just Want to Have Fun," in Lisa A. Lewis, ed., *The Adoring Audience: Fan Culture and Popular Media* (London: Routledge, 1992), 85, 97.

27. Kane, *Ticket to Ride*, 39.

28. Emerick, *Here, There and Everywhere*, 65, 67.

29. Devin McKinney, *Magic Circles: The Beatles in Dream and History* (Cambridge: Harvard University Press, 2003), 159.

30. Ehrenreich, Hess, and Jacobs, "Girls Just Want to Have Fun," 105; *Melody Maker*, 15 Feb. 1964, in Sandercombe, *Press Reports*, 46.

31. Ehrenreich, Hess, and Jacobs, "Girls Just Want to Have Fun," 103.

32. Marcy Lanza quoted in Stark, *Meet the Beatles*, 153.

33. Barbara Boggiano quoted in Berman, *See the Beatles*, 155.

34. Karen Derryberry post in www.thirteen.org/beatles/the-beatles; Interview with Amanda Cody, May 2009; Valerie Volpolni quoted in Berman, *See the Beatles*, 101.

35. Stark, *Meet the Beatles*, 65; Barbara Allen quoted in Berman, *See the Beatles*, 146.

36. O'Brien, "Seven Fat Years," 170.

37. Harold Montgomery quoted in Berman, *See the Beatles*, 203; Michele Hush post in www.thirteen.org/beatles/the-beatles.

38. Kane, *Ticket to Ride*, 73.

39. O'Brien, "Seven Fat Years," 169; Wyn Cooper, "Girls Screaming," in Sawyers, *Read the Beatles*, 301.

40. Ehrenreich, Hess, and Jacobs, "Girls Just Want to Have Fun," 99.

CHAPTER 9. CONVERGENCE

1. Gould, *Can't Buy Me Love*, 162.

2. John Lennon *Playboy* interview, Jan. 1981, in Sawyers, *Read the Beatles*, 215.

3. Stark, *Meet the Beatles*, 17; Gould, *Can't Buy Me Love*, 106.

4. *Daily Mirror*, 2 Oct. 1963, 9.

5. Dave McAleer, "The Birth of Rock-n-Roll," Davemcaleer.com/page13.htm; Eric Hobsbawn, *The Age of Extremes* (New York: Vintage, 1995), 328.

6. Stark, *Meet the Beatles*, 139; Dominic Sandbrook, *White Heat: A History of Britain in the Swinging Sixties* (London: Abacus, 2007), 104; *Daily Mirror*, 2 Oct. 1963, 9.

7. Neaverson, *Beatles Movies*, 81.

8. Ibid., 3.

9. *Newsweek*, 24 Feb. 1964, 54–57; *New York Times*, 17 Feb. 1964, 1, 20.

10. Penny Wagner quoted in Berman, *See the Beatles*, 92; Frontani, *Image and Media*, 92.

11. *Daily Mirror*, 22 Oct. 1963, 16–17.

12. Billy J. Kramer, "The British Invasion," presented at the Museum of Television and Radio Seminar, New York City, 1989.

13. The audience made up about 7 percent of the entire population. Howard Kramer, "Rock and Roll Music," in Kenneth Womack, ed., *The Cambridge Companion to the Beatles* (London: Cambridge, 2009), 69.

14. Paul Harris, *When Pirates Ruled the Waves* (London: Kennedy and Boyd, 2007), 232.

15. Schwensen, *Beatles in Cleveland*, 109; Kane, *Ticket to Ride*, 20.

16. Spizer, *Beatlemania*, 85.

17. Barnouw, *Tube of Plenty*, 314, 331.

18. *American Bandstand* was taped weeks, sometimes months, ahead of its broadcast date; John A. Jackson, *American Bandstand: Dick Clark and the Making of a Rock'n'Roll Empire* (New York: Oxford University Press, 1997), 241.

19. Mark Lewisohn, *The Complete Beatles Chronicle* (London: Pyramid, 1992), 355.

20. *Anthology*, 119; Miles, *Diary*, 106.

21. Blaney, *Beatles for Sale*, 106.

22. Tony Gray, *Fleet Street Remembrances* (London: Heinemann, 1996), 61–62.

23. Dennis Griffiths, *Fleet Street: 500 Years of the Press* (London: British Library, 2000), 327–28.

24. Bramwell, *Magical Mystery Tours*, 102–3. Maureen O'Grady of *Boyfriend* was another journalist close to the Beatles; Norman, *John Lennon*, 299.

25. Sandbrook, *White Heat*, 104.

26. Jacques Attali, *Noise: The Political Economy of Music* (Minneapolis: University of Minnesota Press, 1985), 103.

27. *Daily Mirror*, 20 March 1964, 18. Harold Wilson's constituency was in Liverpool, and he took special care to represent himself as a man of the people.

28. Barbara Allen quoted in Berman, *See the Beatles*, 48.

29. Greil Marcus, "Another Version of the Chair," in Sawyers, *Read the Beatles*, 81–82.

30. Lester Bangs, *Psychotic Reactions and Carburetor Dung* (New York: Vintage, 1988), 325; Wilde, "McCartney: Life in the Shadow," in Sawyers, *Read the Beatles*, 247.

CHAPTER 10. TECHNOLOGY

1. David J. Apple and John Sims, "Harold Ridley and the Invention of the Intraocular Lens," *Survey of Ophthalmology* 40, no. 4 (Jan./Feb. 1996): 285.

2. David Beaty, *The Water Jump: The Story of Transatlantic Flight* (New York: Harper and Row, 1976), 200, 226, 267.

3. Spizer, *Beatlemania*, 82–83.

4. Fred Taylor, ed., *The Goebbels Diaries 1939–1941* (New York: Putnam's Sons, 1983), 183.

5. *Anthology*, 80–81; Mo Foster, *17 Watts? The Birth of British Rock Guitar* (London: MPG Books, 1997), 138–43.

6. *Newsweek*, 18 Nov. 1963, 104; Norman, Lennon, 290. The Beatles came back from their first American tour dissatisfied with the insufficient volume of AC 30s, and Vox built them 50 watt and finally 100 watt amps for their tours; Babiuk, *Beatles Gear*, 103. It was only at the end of the 1960s and the introduction of Marshall's 100 watt amps combined with banks of 4 × 12 inch speaker cabinets that the band on the stage could overcome the noise generated by the audience.

7. Frame, *Restless Generation*, 141.

8. Goldsmith, *Come to America*, 20.

9. André Millard, *America on Record: A History of Recorded Sound* (New York: Cambridge University Press, 2005), 223–24.

10. Spitz, *The Beatles*, 121.

11. Spitz, *The Beatles*, 493; *Mersey Beat*, 30 July 1964, *Imagine* (DVD); *Record Mirror*, 15 May 1965, in Sandercombe, *Press Reports*, 77, 126. The 9 inch Sony television shown on the cover of *Sgt Pepper* was purchased by Paul during a tour of Japan, "Beatles 'Sgt Pepper' Album Cover Mystery a Piece of Japanese History," *Japan Times*, 10 July 2010, www.japantimes.co.jp/cgi-bin/nn21000710cc.html.

12. *Disc Weekly*, 19 Feb. 1956, in Sandercombe, *Press Reports*, 156.

13. Mark Haywood, *The Beatles Unseen* (London: Weidenfeld and Nicolson, 2005), 7; Robert Freeman, *Yesterday: Photographs of the Beatles* (New York: Thunder's Mouth, 1983), 11.

14. *Daily Mirror*, 16 Nov. 1963, 1.

15. Brown, *The Love You Make*, 251. When the Beatles arrived at the ashram, they learned that the maharishi also planned to make a documentary of them and had already sold the rights to American ABC television!

16. Like most of their guitar-playing peers, the Beatles' technological enthusiasm did not extend to transistorized amplifiers, and they kept to tube amps, refusing new solid-state amplifiers offered to them by the Rickenbacker company; Babiuk, *Beatles Gear*, 108.

17. Pete Kennedy quoted in Berman, *See the Beatles*, 37.

18. Gould, *Can't Buy Me Love*, 182.

19. Anna McCormack and Penny Wagner quoted in Berman, *See the Beatles*, 42, 52. By comparison, a console hi-fi system with radio and wood grain finish could cost $500.

20. Shaun Weis quoted in ibid., 65.

21. Jerry Bishop quoted in Schwensen, *Beatles in Cleveland*, 59.

22. Kane, *Ticket to Ride*, 66–67.

23. Womack, *Long and Winding Roads*, 3–4.

24. Bramwell, *Magical Mystery Tours*, 27.

25. *Daily Mail*, 23 Oct. 1963, 8.

26. *Melody Maker*, 16 July 1966, in Sandercombe, *Press Reports*, 171.

27. *Anthology*, 77.

28. The Beatles got only one disc out of this transaction, and it was duly passed around to each member to enjoy; Womack, *Long and Winding Roads*, 15.

CHAPTER 11. IN THE RECORDING STUDIO

1. Paul Saltzman, *Beatle Symposium, Stereo Review* 2 (1973), cited in Stark, *Meet the Beatles*, 182; Brian Epstein interview in *Melody Maker*, 19 Aug. 1967, in Sandercombe, *Press Reports*, 203; Kevin Minaham quoted in www.thirteen.org/beatles/the-beatles.

2. Lennon told Maureen Cleave, "None of my gadgets really work except the gorilla suit"; "How Does a Beatle Live: John Lennon Lives Like This," *Evening Standard*, 4 March 1966, in Sawyers, *Read the Beatles*, 89; Geoff Emerick's views of their technical knowledge is in *Here, There and Everywhere*, 139, 151.

3. Peter Doyle, *Echo & Reverb: Fabricating Space in Popular Music Recording* (Middletown, CT: Wesleyan, 2005), 184.

4. *Anthology*, 92; *Daily Mirror*, 22 Oct. 1963, 16–17.

5. Gillett, *Sound of the City*, 264.

6. Southall, *Abbey Road*, 57; Martin, *All You Need Is Ears*, 144–45.

7. Emerick, *Here, There and Everywhere*, 142.

8. EMI adopted new technology but was conservative when it came to incorporating it in the work process, because every new machine it built or acquired spent months in testing before it was eventually put into use; ibid., 199–200.

9. Emerick, *Here, There and Everywhere*, 54; Glyn Johns, "On Recording the Let It Be Album," in John Tobler and Stuart Grundy, *The Record Producers*, in Egan, *Mammoth Beatles*, 225; *Mersey Beat*, 3 Jan. 1963, 4.

10. Howlett, *The Beatles at the Beeb*, 17.

11. Emerick, *Here, There and Everywhere*, 45.

12. Johnny Gustafason quoted in Geller and Wall, *Epstein Story*, 55; Oriole put strings on a Blue Mountain Boys recording of a Hank Thompson song; McManus, *Nashville of the North*, 13.

13. Emerick, *Here, There and Everywhere*, 42, 58.

14. *Anthology*, 75.

15. *Mersey Beat*, 3 Jan. 1963, 4.

16. Womack, *Long and Winding Roads*, 82.

17. Blaney, *Beatles for Sale*, 72; *Anthology*, 206.

18. Lewisohn, *Beatles Recording Sessions*, 36.

19. Ibid., 64

20. Ian MacDonald, *Revolution in the Head: The Beatles' Records and the Sixties* (London: Pimlico, 1995), 108.

21. Lewisohn, *Beatles Recording Sessions*, 8, 54. Paul McCartney told Lewisohn that the late-night schedule gave the band the freedom to use the entire facility at will, and with no interruptions.

22. Ibid., 70, quoting Jerry Boys; Emerick, *Here, There and Everywhere*, 91.

23. Emerick, *Here, There and Everywhere*, 8–9.

24. Lewisohn, *Beatles Recording Sessions*, 99; Martin, *All You Need Is Ears*, 141.

25. Emerick, *Here, There and Everywhere*, 110–13.

26. Andrew Loog Oldham, *Stoned: A Memoir of London in the 1960s* (New York: St Martin's, 2001), 253, argues that using small studios produced the unique "feel" of the early Rolling Stones records.

27. Emerick, *Here, There and Everywhere*, 298, 302.

28. *Melody Maker*, 27 Feb. 1965, in Sandercombe, *Press Reports*, 114.

CHAPTER 12. THE BEATLES AND THE SIXTIES

1. Touré, "Too Late for the Show," in Sawyers, *Read the Beatles*, 333.

2. Goldsmith, *Come to America*, 168; *Disc and Music Echo*, 11 Nov. 1967, in Sandercombe, *Press Reports*, 218.

3. *The Guardian*, 16 Nov. 2010.

4. Russell Richey interview, Oct. 2010.

5. Stephen Fisch posting on www.thirteen.org/beatles/the-beatles.

6. Bangs, "Withering Away," in Egan, *Mammoth Beatles*, 355; *New York Times* magazine, 16 Aug. 2009, 35.

7. Postings on www.thirteen.org/beatles/the-beatles. Debbie Leavitt can claim to be the first Beatles fan in the United States because she got singles mailed to her by a penpal in Liverpool. Of her experience as a fan and meeting the band, she said afterward, "Thank God I was born at the right time"; Berman, *See the Beatles*, 31, 272.

8. Jon Wilde, "McCartney: My Life in the Shadow of the Beatles," in Sawyers, *Read the Beatles*, 247.

9. Ibid., 248.

10. Larry Johnson, Leslie Healey postings on www.thirteen.org/beatles/the-beatles.

11. Frontani, *Image and Media*, 14; *Anthology*, 153.

12. Kane, *Ticket to Ride*, 108.

13. "Communism, Hypnotism and the Beatles" pamphlet, 1965, quoted in J. Hoberman, *The Dream Life: Movies, Media and the Mythology of the Sixties* (New York: New Press, 2003), 140; Ken King, paper in British Invasion Seminar, Birmingham, Alabama, Aug. 2003; Frontani, *Image and Media*, 8, 40.

14. Posting on www.thirteen.org/beatles/the-beatles; O'Brian, "Seven Fat Years," in Sawyers, *Read the Beatles*, 171.

15. Christopher Porterfield in *Time*, 22 Sept. 1967, 62.

16. Frontani, *Image and Media*, 123–24.

17. Sandbrook, *White Heat*, 210.

18. Carol Cox quoted in Berman, *See the Beatles*, 75.

19. Goldsmith, *Come to America*, 34.

20. Ibid., 141.

21. Geller and Wall, *Epstein Story*, 35.

22. Brown, *The Love You Make*, 142.

23. Postings on www.thirteen.org/beatles/the-beatles; Goldsmith, *Come to America*, 5.

24. Robin Hallum-Goldsmith, Rick Albert, and Darren Divivo postings on www.thirteen.org/beatles/the-beatles.

25. Wenner, *Lennon Remembers*, 12.

26. *How the Beatles Rocked the Kremlin* (Blakeway Productions/WNET, 2009).

27. Spitz, *The Beatles*, 397, 418. Broadway actors demonstrated and petitioned the president to stop giving work permits to British players who were taking their jobs; *Daily Mail*, 12 Oct. 1963, 1.

28. Spizer, *Beatlemania*, iv. The Beatles' American tours were booked by the General Artists Corporation, and some of their people flew back and forth to London to set up the arrangements with NEMS.

29. Sandbrook, *White Heat*, 248. Vidal Sassoon went to cut hair in the United States in 1965, including Mia Farrow's for the film *Rosemary's Baby*, a famous cut that became the model for millions of American women. He said, "We worked very hard to make British hairdressing the best in the world," *Times* (London), 21 Oct. 2009, T2, 2.

30. Sandbrook, *White Heat*, 231.

31. *Observer*, 6 June 2010.

32. Geller and Wall, *Epstein Story*, 106.

33. Lennon quote in *Disc and Music Echo*, 30 March 1968, in Sandercombe, *Press Reports*, 237.

34. Kane, *Ticket to Ride*, 119: Shaun Weiss and Penny Wagner quoted in Berman, *See the Beatles*, 60, 129; Ralph Soloman posting on www.thirteen.org/beatles/the-beatles.

35. Dale Ford quoted in Berman, *See the Beatles*, 114.

36. Wilde, "McCartney: My Life in the Shadow," in Sawyers, *Read the Beatles*, 246.

37. Shaun Weiss quoted in Berman, *See the Beatles*, 71.

38. Ron Sweed quoted in Schwensen, *Beatles in Cleveland*, 57.

39. Charlie Manson was not inspired to commit horrible crimes as a consequence of listening to Beatles' songs like "Helter Skelter"; rather he thought that the band was confirming his own belief that a race war was about to erupt in America. See Vincent Bugliosi, *Helter Skelter: The True Story of the Manson Murders* (New York: Norton, 1974).

40. Geri Montefusco posting on www.thirteen.org/beatles/the-beatles.

41. Cooper, "Girls Screaming," 301; Anthony DeCurtis, "Crossing the Line: The Beatles in My Life," in Sawyers, *Read the Beatles*, 302.

42. Frontani, *Image and Media*, 44–45; *New York Times*, 1 Dec. 1963, 124; Stark, *Meet the Beatles*, quote under image opposite 152.

Index